Research and Practice in Chemistry Education

Madeleine Schultz · Siegbert Schmid ·
Gwendolyn A. Lawrie
Editors

Research and Practice in Chemistry Education

Advances from the 25th IUPAC International Conference on Chemistry Education 2018

Springer

Editors
Madeleine Schultz
Deakin University
Geelong, VIC, Australia

Siegbert Schmid
The University of Sydney
Sydney, NSW, Australia

Gwendolyn A. Lawrie
University of Queensland
St Lucia, QLD, Australia

ISBN 978-981-13-6997-1 ISBN 978-981-13-6998-8 (eBook)
https://doi.org/10.1007/978-981-13-6998-8

Library of Congress Control Number: 2019933387

This Springer imprint is published by the registered company Springer Nature Singapore Pte Ltd.
The registered company address is: 152 Beach Road, #21-01/04 Gateway East, Singapore 189721, Singapore

Preface

Attending the biennial IUPAC International Conference on Chemistry Education (ICCE) is always a rewarding experience that permits dialogue between people working across cultural and geographic distances. The attendees, from over 30 different countries, share a goal of improving student understanding of chemistry. Among the interesting aspects of attendance is learning how educators within very different countries and contexts have encountered similar challenges in teaching chemistry, and their approaches to solving these.

The chapters in this book sample some of the research that was disseminated at ICCE 2018 in Sydney, Australia (the 25th iteration of this meeting). They are arranged according to four themes that emerged through the conference:

- Promoting undergraduates' understanding of chemistry through novel pedagogies and innovative strategies in new learning environments.
- Responding to the dynamic nature of teaching environments.
- Translating assessment into the next dimension.
- Developing professional skills in chemistry students.

These themes combine to present a landscape of current issues facing chemistry educators due to the increasing integration of technology into teaching, increasing class sizes, institutional and regulatory interest in the evaluation of students, and the changing role of universities in providing vocational skills. Strategies developed by researchers, working within their own specific contexts, are frequently more widely applicable, and in many cases, readers of this book will find the background narrative to the research compelling, even if the outcomes are not directly relevant to their own work. Together, the chapters provide the reader with a diverse set of reports covering both theoretical and practical aspects of chemistry education at different levels.

It is interesting to observe both similarities and changes in emphasis within chemistry education that have occurred during the past decade as illustrated by a comparison of this current book with two previous volumes sharing contributions from the 2008 (20th ICCE) and 2010 (21st ICCE) iterations of this conference (also published by Springer). Those meetings were structured around overarching themes of ICT and Sustainability in 2008 and 2010, respectively. In contrast to 2018, while these themes remained visible, approaches to integrating them into teaching practice have matured, with many technologies becoming commonplace in the intervening years. The 2018 ICCE conference had a greater emphasis on educational theory and its application within teaching practice, including professional development of chemistry educators. I hope that you enjoy reading the research outcomes and the teaching strategies presented here and possibly find inspiration to change your own practice.

I thank Gwen and Siggi for their editorial help throughout the process of preparing this book. I am grateful to all chapter authors for their contributions, to our reviewers, and to Springer for support during the preparation of the book.

Geelong, VIC, Australia
January 2019

Madeleine Schultz

Contents

Chapter 1
Introduction

Gwendolyn A. Lawrie

Abstract The biennial IUPAC International Conference on Chemistry Education (ICCE) provides participants with an opportunity to share their research and/or their practices or to simply survey the landscape in this rich field and exchange ideas with peers. Participants typically represent multiple academic and teaching roles at all career stages. The international roots of this forum provide us with a snapshot of chemistry education research and practice originating across all five continents—we are able to map a landscape that informs alignment with the current understanding and the directions in growth in our field.

The biennial IUPAC International Conference on Chemistry Education (ICCE) provides participants with an opportunity to share their research and/or their practices or to simply survey the landscape in this rich field and exchange ideas with peers. Participants typically represent multiple academic and teaching roles at all career stages. The international roots of this forum provide us with a snapshot of chemistry education research and practice originating across all five continents—we are able to map a landscape that informs alignment with the current understanding and the directions in growth in our field. Fourteen studies that were shared at ICCE 2018 (marking the 25th anniversary of this meeting) have been assembled as chapters in this book to provide the reader with insights from international contexts and perspectives in the field. The authors represent the diversity in educational contexts and cultures that ICCE celebrates, and the chapters are arranged according to four themes that emerged through the conference:

- Promoting undergraduates' understanding of chemistry through novel pedagogies and innovative strategies in new learning environments.
- Responding to the dynamic nature of teaching environments.
- Translating assessment into the next dimension.
- Developing professional skills in chemistry students.

G. A. Lawrie (✉)
University of Queensland, St. Lucia, QLD, Australia
e-mail: g.lawrie@uq.edu.au

M. Schultz et al. (eds.), *Research and Practice in Chemistry Education*,
https://doi.org/10.1007/978-981-13-6998-8_1

However, these chapters could easily have been clustered according to other traditional categories, for example, students' conceptions, technology in education, visualisation and communication skills. The aim of this book is to capture advances from the different lenses of chemistry education: pedagogies and practices; students' learning environments; measuring learning; and articulating students along career pathways.

Promoting Undergraduates' Understanding of Chemistry Through Novel Pedagogies and Innovative Strategies in New Learning Environments

A teacher's work involves knowing their students and how they learn, including their prior experiences, their current knowledge, and what they will learn in the future—this represents teacher professional knowledge and skills (Gess-Newsome, 2015). A core part of this endeavour is in engaging students in the processes of developing or sharing their mental models of chemical concepts. The pedagogies and practices adopted by chemistry teachers are often informed by the enormous collective knowledge base that has been assembled during decades of education research – dramatic, revolutionary shifts in practice are rarely observed. Therefore, conference programs often include shared experiences, frequently empirical in nature, that are deeply contextualised and incremental in findings. To encourage the shift towards data-driven practice, three chapters that give exemplars of this have been included in this book as exemplars of innovative pedagogies and practices demonstrated in hybrid learning environments.

In recent years, hybrid learning environments have become almost ubiquitous through adoption of computers and associated technologies to enhance learning. They play an integral role in delivering multimodal representations that aim to engage students in visualisation of chemical phenomena including dynamic animations and simulations as well as the realms of virtual reality. Akaygun and Adadan (Chap. 2) challenge pre-service teachers to critique and reflect on their own thinking after viewing and generating storyboards of correct and incorrect animations of redox reactions. Student-generated contexts can offer insights into their thinking. In Chap. 3, Paderna et al. developed a teaching approach that requires students to generate examples of application of chemistry concepts and to present these. Immersive virtual reality experiences represent the cutting-edge of engaging students in testing their mental models of phenomena involving the application of their visuospatial skills. Substantial development is still required to demonstrate the benefits of these learning experiences, and the work reported by Won et al. (Chap. 4) is a good example of establishing the factors that may need to be explored.

Responding to the Dynamic Nature of Teaching Environments

Teachers work within fluid and ever-evolving environments; they are encouraged to constantly reflect on student learning in terms of what the student does (Biggs, 1999; Biggs & Tang, 2011), while also evaluating their own practices (Kane, Sandretto, & Heath, 2004). There is no doubt that instructional design and the desire to introduce adaptive learning within active and blended learning environments require orchestration of activities using increasingly diverse resources and tools (Dillenbourg, 2013).

Shared reflections provide valuable insights for other teachers and enable the community to develop practices that address difficult challenges. In this section of the book, three chapters provide a lens on diverse challenges, and successes, in teachers' work including motivating faculty in historically underrepresented groups in adoption of high-impact, evidence-based pedagogies (Wilson-Kennedy et al., Chap. 5). The shift of faculty into blended-learning practices catalysed by external disruption and impact on student learning (Potgieter et al., Chap. 6) provides insights through a shared narrative of teachers through this journey. Finally, an example of reconsidering the philosophical basis of thinking processes across multiple levels in science to reconsider approaches in evaluating how students develop their thinking is presented (Graulich & Casperi, Chap. 7).

Translating Assessment into the Next Dimension

Assessment of student knowledge and skills is crucial for measuring their learning outcomes as feedback to student, teacher, institution and external stakeholders (NRC, 2013; Brown, Bull, & Pendlebury, 2013). The nature of learning actually measured, and its value, often receives too little attention when chemistry teachers rely upon traditional modes of assessment such as examinations, quizzes and in particular laboratory activities across both secondary and tertiary contexts (Hofstein & Lunetta, 2004). There are now widespread examples of research-informed studies in chemistry education, founded on strong theoretical or conceptual frameworks, that share data collected in a new ways from new contexts that add to the collective body of knowledge and practices in assessing learning. Three chapters are included in this book that highlight three new directions in assessment. In the first (Chap. 8), Talanquer presents Chemical Thinking and new approaches to including formal formative assessment and summative assessment of students' abilities to integrate central ideas and apply cross-cutting ways of reasoning in chemistry. Analysis of examination items to identify the type of learning that is being measured is considered by Erduran et al., (Chap. 9) with the aim of shifting the focus of assessment to practical science. Tan et al., (Chap. 10) discuss the affordances, including versatility, of using online data collection to investigate students' concepts of ionisation energies. This further reinforces the affordances that technology-enhanced assessment offers to instructors in terms of formative feedback.

Developing Professional Skills in Chemistry Students

Inspection of issues by any leading journals that publish chemistry education research and practice from the past 5 years will reveal the increased number of articles that focus on developing students' skills beyond disciplinary knowledge and practice in the curriculum. Indeed, a special issue of Chemistry Education Research and Practice (issue 3, volume 18, 2017) was themed 'Development of key skills and attributes in chemistry' recognising that chemistry education can contribute to broader development of students (Taber, 2016). There are multiple terms used to describe these skills: transferable, generic, soft, employability skills, etc. Regardless of the terminology, they all represent professional skills that employers value and have been historically been neglected in tertiary studies. The last section of this book focuses on practices and assessment that support students and faculty in their awareness and development of professional skills. There are five chapters in this section, emphasising the contemporary shift within our field towards this dimension in learning in chemistry.

Students' recognition of the importance of professional skills can be developed in tailored programs. Pugh (Chap. 11) describes a strategic initiative enabling chemistry students to recognise and develop business skills, while McHugh et al. (Chap. 12) focus on developing a specific skill set related to communication and public engagement through learning modules. Indeed communication skills are often espoused as a learning outcome from collaborative group work and new approaches to measuring students' gains in these (along with other professional skills) in active learning environments form the basis of Chaps. 13–15. Recognising the important foundation of constructive alignment in instructional design (Biggs, 1996), Cole et al. present several case studies demonstrating explicit assessment of professional skills using innovative rubrics and a variety of instructional practices (Chap. 13). Yeung and Ahern (Chap. 14) describe an intervention in practice to improve communication between individual students in group tasks in the classroom. The final Chap. 15 (Schultz et al.,) in this book presents an evaluative tool that teachers can adopt as part of curriculum mapping to demonstrate to what extent their assessment actually provides students with opportunity to demonstrate proficiency in transferable skills in chemistry courses or programs of study.

Summary

Recent reviews of science and chemistry education research and practice reinforce the strong foundations of knowledge and theory that our community builds upon. Cooper and Stowe (2018) draw together the numerous threads in chemistry education research to pose three questions: (1) What should students know and be able to do with that knowledge? (2) How will we know that students have developed a coherent and useful understanding of chemistry? (3) What evidence do we have about how to help students develop a deep and robust understanding of chemistry? The chapters shared in this book indicate that the wider community implicitly hold these important questions as central to their own endeavours. However, they are also examining wider horizons in terms of how best to support teaching practices and the articulation of students into future careers.

Acknowledgements The editorial team is very grateful to the many reviewers who reviewed chapters for this book. We also thank all chapter authors for their contributions.

References

Biggs, J. (1996). Enhancing teaching through constructive alignment. *Higher Education, 32,* 347–364. https://doi.org/10.1007/BF00138871.

Biggs, J. (1999). What the student does: Teaching for enhanced learning. *Higher Education Research & Development, 18,* 57–75. https://doi.org/10.1080/0729436990180105.

Biggs, J., & Tang, C. (2011). *Teaching for Quality Learning at University: What the Student Does* (4th ed.). Maidenhead: SRHE and Open University Press.

Brown, G. A., Bull, J., & Pendlebury, M. (2013). Assessing student learning in higher education. Routledge.

Cooper, M. M., & Stowe, R. L. (2018). Chemistry education research—from personal empiricism to evidence, theory and informed practice. *Chemical Reviews, 118*(12), 6053–6087. https://doi.org/10.1021/acs.chemrev.8b00020.

Dillenbourg, P. (2013). Design for classroom orchestration. *Computers & Education, 69,* 485–492. https://doi.org/10.1016/j.compedu.2013.04.013.

Gess-Newsome, J. (2015). A model of teacher professional knowledge and skill including PCK: Results of the thinking from the PCK Summit. In A. Berry, P. Friedrichsen, & J. Loughran (Eds.), *Re-examining pedagogical content knowledge in science education* (pp. 28–42). New York, NY: Routledge.

Hofstein, A., & Lunetta, V. N. (2004). The laboratory in science education: Foundations for the twenty-first century. *Science education, 88*(1), 28–54. https://doi.org/10.1002/sce.10106.

Kane, R., Sandretto, S., & Heath, C. (2004). An investigation into excellent tertiary teaching: Emphasising reflective practice. *Higher Education, 47,* 283–310. https://doi.org/10.1023/B:HIGH.0000016442.55338.24.

NRC. (2013). *Developing Assessments for the Next Generation Science Standards.* Washington, DC: The National Research Council, National Academies Press.

Taber, K. S. (2016). Learning generic skills through chemistry education. *Chemistry Education Research & Practice, 17,* 225–228. https://doi.org/10.1039/c6rp90003h.

Chapter 2
Revisiting the Understanding of Redox Reactions Through Critiquing Animations in Variance

Sevil Akaygun and Emine Adadan

Abstract Animations and simulations are frequently used to teach concepts of chemistry, especially at the submicroscopic level. They often help students visualize the processes occurring among the species, but students may not necessarily change their mental models, after viewing them. In order to elicit their understandings, it may be suggested to allow students not only to passively view but also to critique the animations. With adopting the variation theory, this study aimed to investigate how preservice chemistry teachers make sense of two dynamic submicroscopic redox animations that are conceptually in conflict with each other by critiquing and reflecting on their particular features. Twenty-two preservice chemistry teachers first viewed an experimental video of a redox reaction between zinc metal and copper(II) sulfate solution. Then, they represented this redox reaction at the submicroscopic level. These representations, revealing their mental models, were coded and analyzed to identify the level of their conceptual understanding. Then, they viewed two submicroscopic animations; one of them was showing a scientific representation, and the other included a common alternative conception. After viewing these animations, the participants wrote reflections by critiquing them. One week later, they were interviewed to obtain further information about their understandings. Findings from the analysis of reflective writings and interviews suggested that preservice chemistry teachers' initial mental models affected the aspects they highlighted in their representations. The participant with a scientific understanding (1 of 22) was able to identify the scientifically accurate animation as accurate. About two-third of the participants with a moderate, a weak and an alternative understanding (13 of 21) identified the scientifically accurate animation as accurate; whereas about one-third of the same group of participants (7 of 21) did not clearly indicate their views about the scientifically accurate animation but they only described its particular features without referring to its accuracy. In general, the majority of participants experienced a challenge with making connections between macroscopic and submicroscopic features.

S. Akaygun (✉) · E. Adadan
Bogazici University, Beşiktaş, Turkey
e-mail: sevil.akaygun@boun.edu.tr

E. Adadan
e-mail: emine.adadan@boun.edu.tr

M. Schultz et al. (eds.), *Research and Practice in Chemistry Education*,
https://doi.org/10.1007/978-981-13-6998-8_2

Introduction

Science, in particular chemistry, is multimodal in nature so that chemists frequently represent chemical phenomena employing the three levels of representations, namely *macroscopic* (tangible, visible), *submicroscopic* (molecules, atoms, ions), and *symbolic* (chemical equations, diagrams) (Cheng & Gilbert, 2009; Johnstone, 1991). Similarly, learning chemistry, particularly chemical reactions, requires students to make sense of observable changes and macroscopic features of phenomenon in connection to the movement and interaction of atoms, molecules, and ions at the submicroscopic level, as well as representing such changes with symbols (Kelly & Hansen, 2017). Research showed that students from middle school to university have challenges visualizing dynamic particulate interactions in chemical processes and smoothly moving and making connections across three levels of representation (Johnstone, 1991; Kozma, 2003; Stieff, 2011; Taber, 2013).

Dynamic visualizations, in other words, animations provide plentiful potential for chemistry learning. Such tools often represent chemical phenomenon in multiple forms and show translations between macroscopic, submicroscopic, and symbolic representations of phenomenon in a coordinated manner (Ainsworth, 1999; Zhang & Linn, 2011). In other words, multiple representations provide students with more comprehensive image of phenomenon, comprising certain pieces of information in each individual representation (Ainsworth, 1999). With dynamic visual tools, students become able to visualize dynamic movement and interactions among invisible entities (atoms, molecules, ions) represented by chemical symbols. Therefore, they more likely develop an integrated understanding of particular chemical phenomenon as they recognize the referential links between its three levels of representation (Ardac & Akaygun, 2004; Kelly, 2014; Seufert, 2003; Tasker & Dalton, 2008).

Dynamic submicroscopic animations are frequently used for instructional purposes to show students certain aspects associated with the particular chemical concept or phenomenon. Findings from several studies indicated that students who simply viewed the dynamic submicroscopic animations improved their conceptual understandings at the submicroscopic level (e.g., Ardac & Akaygun, 2004; Chang & Linn, 2013; Kelly, 2017; Williamson & Abraham, 1995; Wu, Krajcik, & Soloway, 2001). Students generally incorporate certain aspects from animations to their conceptual framework, when they just view dynamic animations. However, they often may not develop full scientific understandings due to confusing features of or cognitive load created by dynamic submicroscopic animations (Kelly, 2014; Lowe, 2003; Rosenthal & Sanger, 2012). Research suggests that effective use of dynamic submicroscopic animations closely rely on proper design and scaffolding of learning (Quintana et al., 2004). For example, Chang and Linn (2013) reported that students who were engaged in critiquing of dynamic animations indicated substantial conceptual progress compared to students who just observed the dynamic animations and those who were offered guidance as they interacted with dynamic animations.

Students often hold flawed or naive ideas prior to instruction, called alternative conceptions, and such ideas are strongly held and usually interfere with further science learning (Smith, di Sessa, & Roschelle, 1993). Based on constructivist view of learning, students' alternative conceptions can be considered as resources for conceptual progression, and consequently, instruction for conceptual change should have potential to create a disequilibrium in relation to students' alternative conceptions, providing them an opportunity to confront with their existing conceptions (Treagust & Duit, 2008). In the same line, when learning from dynamic submicroscopic animations is considered, Plass, Homer, and Hayward, (2009) stated that the efficacy of such tools "depends on whether learners have sufficient cognitive resources to perceive and process the essential information in dynamic visualizations" (p. 33). Research showed that regardless of adequacy of their cognitive resources, when students just view scientifically accurate dynamic submicroscopic animations, they usually focus on its general characteristics and spend less attention to the detailed features it includes, and so they even dismiss some of its important features as unimportant (Kelly, 2014). However, when students compare and contrast dynamic submicroscopic animations of varying accuracy (scientific and nonscientific), they thoroughly think about the accuracy of animations and their consistency with experimental evidence, as well as integrating more scientific features from animations into their conceptual frameworks (Kelly, Akaygun, Hansen, & Villalta-Cerdas, 2017). This finding indicates the importance of offering students an opportunity to critique dynamic submicroscopic animations rather than just having them passively view such animations. In this respect, the purpose of the study was to investigate how pre-service chemistry teachers critique and respond to two submicroscopic animations of a redox reaction in which one shows a scientifically acceptable process (*accurate*) consistent with experimental evidence and another one shows an alternative representation (*inaccurate*) of the same reaction process. Subsequently, the study also examined which animation participants consider as more scientific representation of a redox reaction.

Theoretical Framework

Chemistry instruction primarily aims to develop students' conceptual understanding in the related domain. In doing so, students are supposed to be able to employ what they have learned in particular contexts, as well as building well-connected and hierarchically organized conceptual frameworks (Stevens, Delgado, & Krajcik, 2010). This is an active mental process that entails interpreting a new concept with respect to already existing conceptions and associating it with an existing conceptual framework (Treagust & Duit, 2008). From a conceptual change perspective, learning either results from the addition of new concepts into an existing conceptual framework (conceptual growth) or the major reorganization of existing knowledge to incorporate new knowledge into the conceptual framework (conceptual change) (Posner, Strike, Hewson, & Gertzog, 1982; Treagust & Duit, 2008). For conceptual change, students should notice the inconsistency between their alternative conception and the newly introduced scientific conception and become dissatisfied with their own conception. Consequently, conceptual competition takes place following the comparative status of intelligibility, plausibility, and fruitfulness of opposing conceptions (Posner et al., 1982). Treagust and Duit (2008) found that even after instruction, students often show limited progress toward a scientific understanding due to the robust and resistant nature of alternative conceptions. Research also revealed that students' alternative conceptions either stay alive along with the scientific ones or merge with certain aspects of the scientific concept, creating a *synthetic model* (Treagust & Duit, 2008; Vosniadou, 1994).

Constructivist view of learning asserts that students makes sense of new experience with respect to their existing conceptions and develop unique understandings of a phenomenon (Driver, Squires, & Wood-Robinson, 1994). In the same line, variation theory intends to explain how individuals see, understand, or experience a given phenomenon in a particular manner (Orgill, 2012). Each individual usually makes sense of the same phenomenon differently; this is because individuals often may not *simultaneously* pay attention to and be aware of the critical aspect(s) of a phenomenon. According to variation theory, individuals recognize a particular aspect of a phenomenon, when they experience variation in elements corresponding to that aspect; thus, individuals create meaning for the phenomenon based in their experience of variation (Bussey, Orgill, & Crippen, 2013). The act of recognizing particular aspects of a phenomenon has to do with three key processes, namely, awareness of differences, being able to discern what the differences are, and being simultaneously aware of diverse critical elements. In the present study, variation theory was adopted in such a way that two contrasting animations showing the mechanism of the same redox reaction were designed, and participants viewed both and critiqued them in terms of their accuracy. Participants ultimately need to connect the certain features from animation to experimental evidence and select the more scientific one between the two.

Students' Understanding of Redox Reactions

Electrochemistry, the branch of chemistry, is viewed to be a fundamental topic of General Chemistry courses and textbooks, which looks at the interaction of electricity and chemical reactions. Due to their potential to generate an electric current, redox reactions are the main focus of electrochemistry (Atkins & Jones, 2010). Students' conceptual difficulties associated with electrochemistry, more specifically redox reactions, has been well-studied (e.g., Acar & Tarhan, 2007; Brandriet & Bretz, 2014; Garnett & Treagust, 1992; Österlund & Ekborg, 2009; Rosenthal & Sanger, 2012; Sanger & Greenbowe, 1997; Schmidt & Volke, 2003). One reason for these difficulties has to do with the frequent use of symbolic representations when teaching chemical reactions, such that students were first offered the definition of these reactions, and then they were expected to write and balance the equations of redox reactions (de Jong, Acampo, &Verdonk, 1995; Schmidt & Volke, 2003). Another reason can be students' confusion with multiple definitions for the processes of redox reactions offered by chemistry teachers and textbooks (Rosenthal & Sanger, 2012; Schmidt & Volke, 2003). The first definition, as offered by Lavoisier, is the oxygen method in which gaining oxygen atoms indicates an oxidation reaction, whereas losing oxygen atoms indicates a reduction reaction. The second definition is the electron method in which when a substance loses electrons, it is oxidized, and when a substance gains electrons, it is reduced. Thus, in a redox reaction, oxidation and reduction occur simultaneously. The third definition is the oxidation number method which involves the changes in the oxidation numbers, such that oxidation numbers simultaneously increase and decrease during the redox reaction. The fourth definition is the hydrogen method in which losing hydrogen atoms shows an oxidation reaction and gaining hydrogen atoms shows a reduction reaction. All these definitions require using symbolic level such that students write equations, assign oxidation numbers, and balance equations. However, applying one or more of these definitions using symbols does not necessarily ensure the understanding of processes of redox reactions as long as there is no connection to dynamic particulate interactions (i.e., atom, ions, electrons, etc.) (Taber, 2013). In this respect, instruction should focus on promoting students' understanding of redox reactions at the submicroscopic level, relating macroscopic changes to the behavior of particles rather than just emphasizing symbolic representations of redox reactions.

Some studies revealed that students view oxidation and reduction as independent reactions, and they encounter challenges with the meaning and assignment of oxidation numbers and the identification of reactants as oxidizing or reducing agents (Garnett & Treagust, 1992; de Jong & Treagust, 2002; Österlund & Ekborg, 2009). Acar and Tarhan (2007), in their study with Grade 11 students, identified some new alternative conceptions about the dynamic processes such as the flow of protons and electrons (e.g., protons and electrons flow in the opposite directions to constitute an electric current) and movement of ions (only negatively charged ions constitute a flow of current in the electrolyte). In a recent study, Bandriet and Bretz (2014) reported various alternative conceptions utilizing the information from the Redox Concept Inventory that they developed by eliciting students' understandings through interviews. One was about the oxidation numbers in which students believed that the charge on Zn, Cu, and $(NO_3)^-$ does not change when the reaction equation given as $Zn(s) + Cu(NO_3)_2\ (aq) \rightarrow Zn(NO_3)_2\ (aq) + Cu(s)$. Another one was about the surface features of a given reaction, for example, students believed that the formation of aluminum oxide ($4Al(s) + 3O_2\ (g) \rightarrow 2Al_2O_3(s)$) is not a redox reaction, it is a combination reaction. Students also found spectator ions in redox reaction as irrelevant since they do not appear in net ionic equations. Moreover, students believed that in a single displacement reaction of Cu(s) and $AgNO_3$ (aq), when Cu^{2+} leaves the solid, two electrons remain in the solid. In general, considering the critical role of students' existing conceptions in their subsequent learning, students with strong alternative conceptions about redox reactions may have difficulties in integrating scientific information that they are introduced during instruction into their conceptual frameworks (de Jong & Treagust, 2002; Treagust & Duit, 2008).

Methodology

This qualitative study, which adopted the variation theory, was designed to explore how preservice chemistry teachers differently perceive and critique two submicroscopic animations of a redox reaction occurring between zinc metal and copper(II) sulfate solution. The way they critique the animations were compared with their initial understanding of redox reactions.

Participants

Twenty-two senior preservice chemistry teachers voluntarily participated in the study through a workshop. The average age of them was 24 and their average GPA was 2.60 out of 4.00. Three of the participants (14%) were male, approximately representing the distribution in the population. The preservice teachers had learnt redox reactions in their General Chemistry II and Analytical Chemistry classes. Prior to the study, ethical approval for working with preservice chemistry teachers on reviewing and discussing the features of animations was obtained from the Institutional Review Board. This approval included permission for data collection in the form of digital text files and voice recordings.

Procedure and Data Collection

This study was conducted as a part of a 3-week workshop entitled "Visualizing Chemical Concepts". Every week, the workshop took about 3 h, yielding a total of 9 h for this study. During the first week of the workshop, preservice chemistry teachers viewed a 4-min video[1] of a redox reaction occurring between zinc metal and copper(II) sulfate solution. The video was recorded and edited by the research team. The participants worked individually in a computer lab on their own pace and were allowed to view the video as many times as they liked. The screenshots of the video and the link to access it are given in Fig. 2.1.

After viewing the video, the participants were asked to write reflections, specifically to point out the features that can be considered as evidence for a redox reaction between zinc metal and copper(II) sulfate solution. Then, the participants generated storyboards which are graphical representations that show the sequential flow of a story like a comic book. These storyboards allowed preservice teachers to visually represent the processes of a redox reaction with a series of drawings and written explanations. Then, the participants were interviewed to better grasp their understandings. The following week, they viewed two dynamic submicroscopic animations of this redox reaction, one of them was showing a scientific representation, and the other

[1] https://www.dropbox.com/s/o62epab4k0dkr06/Redoks_Video_06112017_ingilizce.avi?dl=0.

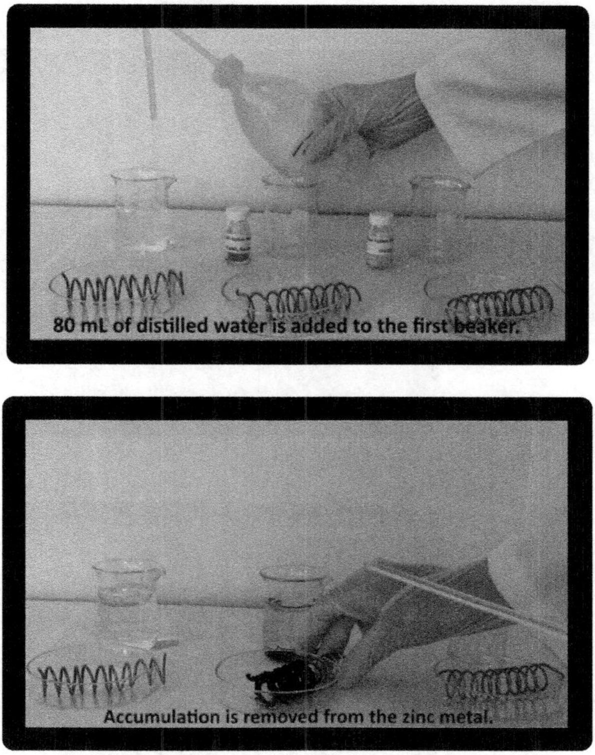

Fig. 2.1 Screenshots of the redox reaction video

included a common alternative conception. The animations were developed by the research team; the scientifically accurate animation was designed by consulting a group of experts from chemistry. The alternative conception as shown in the inaccurate animation was previously determined in the pilot study conducted with another group of preservice chemistry teachers before developing the animations. The participants of the pilot study were another group of 13 senior preservice chemistry teachers; in other words, they were not the participants of the current study. They took an open-ended questionnaire asking the written and pictorial description of a redox reaction at macroscopic, symbolic, and submicroscopic level. Then 12 of the preservice teachers were interviewed to get a deeper understanding of how they conceptualized the redox reactions. After identifying a common alternative conception, which was the deposition of spectator ions on the metal to form a precipitate, an inaccurate animation was developed besides the accurate animation. Specifically, the features of the accurate animation included the motion of copper(II) and sulfate ions, electron transfer from zinc metal to copper(II) ions, formation and releasing of zinc ions into the solution and deposition of copper atoms on top of zinc atoms. The features of inaccurate animation included the motion of copper(II) sulfate as a

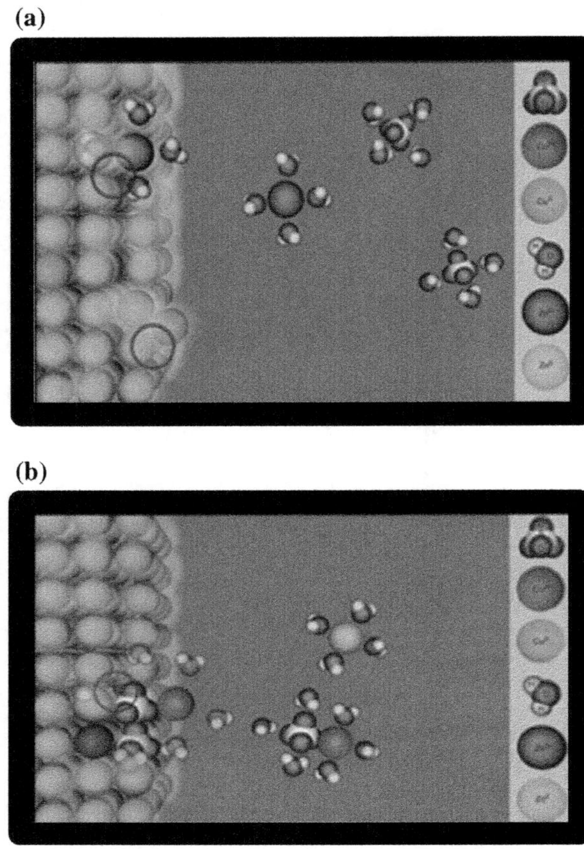

Fig. 2.2 Screenshots of the submicroscopic animations of the redox reaction; **a** scientifically accurate one (https://www.dropbox.com/s/xtci2t42ujbvyr4/Redox_Animation_Accurate.mov?dl=0), **b** the one including an alternative conception (https://www.dropbox.com/s/cnhktut6ymodj9g/Redox_Animation_Inccurate.mov?dl=0)

molecule, electron transfer from zinc atoms to sulfate ions, formation and deposition of zinc sulfate precipitate on the zinc atoms, and releasing of copper atoms into the solution. The screenshots of the animations and the links to access them are given in Fig. 2.2.

After the preservice teachers viewed the animations, they reflected on the features of animations that were justified or refuted by experimental evidence, as well as criticizing the strong and weak aspects of the animations. Within a week, they were interviewed one more time to further capture their understanding about the animations. During the interviews, they were asked to discuss what the animations were depicting, the features that were justified by the experimental evidence, strengths and weaknesses of the animations, and which of the animations better fit to their mental models.

Data Analysis

Data analysis was conducted at two stages. In the first stage, the storyboards generated by the participants were coded, and then the themes and categories emerged from data were identified using constant comparative method (Glaser 1965). Accordingly, storyboards were coded with respect to the criteria as seen in Table 2.1, revealing four different levels of understanding. These representations were compared with the interview transcriptions to check if there were any inconsistencies in the coding. Finally, the levels of understanding categories that emerged from data were identified as a *scientific understanding, a moderate understanding, a weak understanding*, and an *alternative understanding*. These levels were specified by certain criteria as shown

Table 2.1 Criteria showing the levels of understanding of redox reactions

Level of understanding (LU)	Criteria
LU 4: Scientific understanding	Including all types of particles (electrons, ions, atoms, and molecules) and properly showing all the changes in particles along the reaction process
	Drawing and explaining the electron transfer between zinc atoms and copper ions
	Drawing and explaining the deposition of copper atoms on zinc metal and the transition of zinc ions into solution
LU 3: Moderate understanding	Not including all types of particles, and showing some changes in particles along the reaction process, but not all of them
	Explaining the transfer of electron between zinc atoms and copper ions but not drawing the process
	Either drawing or explaining the deposition of copper atoms on zinc metal and the transition of zinc ions into solution
LU 2: Weak understanding	Not including all types of particles, and not showing changes in particles along the reaction process
	Drawing the deposition of copper atoms on zinc metal along with displacement between copper ions and zinc atoms
	Not referring to electron transfer between zinc and copper ions as part of the reaction process but not conveying any alternative understanding
LU 1: Alternative understanding	In addition to weak understanding criteria, the representations included one or more of the following alternative conceptions:
	Drawing copper(II) sulfate in water as molecules not as ions
	Drawing zinc sulfate as molecular precipitate, that is to say, identifying and indicating solid deposition in solution as the precipitation of zinc sulfate
	Not drawing zinc ions going into solution
	Not discriminating atoms and ions involving reaction process

in Table 2.1. For instance, the storyboard of a preservice teacher, shown in Fig. 2.3, was coded as a *weak understanding* (LU 2) because this person did not include water molecules and did not show the changes in particles along the reaction process. The drawing included the deposition of copper atoms on zinc metal but she did not refer to electron transfer between zinc atoms and copper ions as part of the reaction process neither in her drawing nor in her written explanation. Yet, she did not convey any alternative understanding. A more detailed description of level of understanding of redox reaction can be found in another work (Akaygun, Adadan, & Kelly, 2019).

In the second stage of the data analysis, qualitative content analysis (Hsieh & Shannon, 2005) was conducted to analyze preservice teachers' reflective writings and interviews about the submicroscopic animations. Through the open coding, themes and categories were identified. These categories were matched with the levels of understanding of redox reactions to identify how preservice chemistry teachers critiqued and made sense of animations in conflict with each other.

Results and Discussion

The storyboards generated by preservice teachers, after viewing an experimental video, were used to identify their levels of understanding of the redox reactions as mentioned above. Findings from the analysis showed that there was one preservice teacher with a scientific understanding, 4 with a moderate, 7 with a weak, and 10 with an alternative understanding. The levels of understanding of preservice teachers were considered during the analysis of how they critiqued the animations.

When the preservice chemistry teachers' reflective writings and interviews were analyzed, they categorized both animations as *accurate*, *inaccurate*, or *just described* without commenting on its accuracy. Tables 2.2 and 2.3 present how preservice chemistry teachers at different levels of understandings categorized the animations.

After Viewing the Accurate Animation

Based on the findings from the analysis of preservice teachers' views about the accurate animation, one of the participants considered this animation as *inaccurate*, 14 of the participants categorized it as *accurate*, 7 of them *just described* and discussed what they have seen in the animation without categorizing it as accurate or inaccurate. The number of participants who categorized the scientifically accurate animation as accurate, inaccurate, or just described at each level of understanding, and the specific features they emphasized are shown in Table 2.2.

When the preservice teachers were specifically asked to discuss the features of the animation, that were justified by the experimental evidence shown in the video they viewed, they equally referred to macroscopic and submicroscopic features. More specifically, even though they mostly referred to an observation (25, in total), such

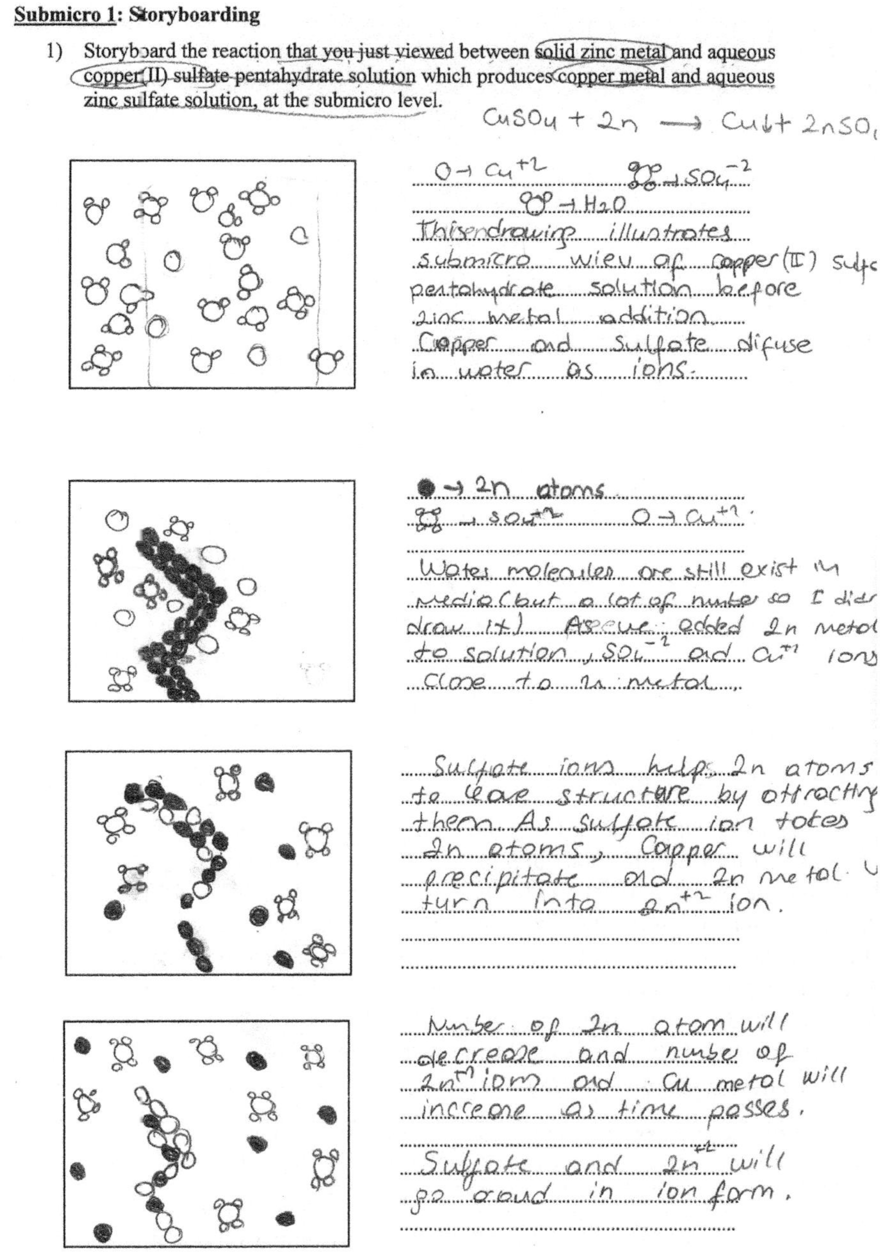

Fig. 2.3 Storyboard of a preservice teacher with a weak understanding

Table 2.2 Preservice teachers' categorization of scientifically accurate animation and the features they mentioned for the justification of experimental evidence for a redox reaction

Level of understanding, LU (N)	Categorization of the scientifically *accurate* animation (N)	Features justified by the experimental evidence (N)
LU 4: Scientific understanding (1)	Accurate (1)	Paling color of the solution (1)
		e-transfer (*submicroscopic*) (1)
		Ionization of Zn^{+2} (*submicroscopic*) (1)
LU 3: Moderate understanding (4)	Accurate (3)	Deposition of Cu (4)
	Just described (1)	Paling color of the solution (2)
		e-transfer (3)
		Ionization of Zn^{+2} (2)
LU 2: Weak understanding (7)	Accurate (5)	Deposition of Cu (6)
	Just described (2)	Paling color of the solution (2)
		e-transfer (4)
		Ionization of Zn^{+2} (5)
LU 1: Alternative understanding (10)	Accurate (5)	Deposition of Cu (8)
	Inaccurate (1)	Paling color of the solution (2)
	Just described (4)	e-transfer (6)
		Ionization of Zn^{+2} (4)

as deposition of copper atoms (18), and paling color of the solution (7), almost equal number of times (26, in total) they referred to the features at the submicroscopic level, such as electron transfer (14), and ionization of zinc atoms (12). Interestingly, this was observed among the preservice teachers regardless of their levels of understanding of redox reactions. For instance, one preservice chemistry teacher, who held a scientific understanding of the redox reactions, in her reflective writing, pointed out an observation concerning the paling color of solution, that was rarely identified among the participants as an indication of a redox reaction. The following quote comes from her reflective writing:

> In the animation, copper ion in the solution is decreasing by time. We have seen [in the video] that blue color of solution (copper ion enables blue color) decreasing which means that the number of copper ions decreasing. (LU 4—accurate)

In her interview, she also included the submicroscopic level features, such that:

> Because there exists electron transfer, ions and metal atoms exchange their places, this is an indication of a redox reaction. (LU 4—accurate)

Some of the preservice teachers with a weak understanding (29%) or an alternative understanding (40%) just described the accurate animation without indicating its accuracy or identifying if it was justified with the experimental evidence. In other

Table 2.3 Preservice teachers' categorization of the inaccurate animation and the features they mentioned for the refusing of experimental evidence for a redox reaction

Level of understanding, LU (N)	Categorization of the scientifically *inaccurate* animation (N)	Features refuted by the experimental evidence (N)
LU 4: Scientific understanding (1)	Inaccurate (1)	Deposition of SO_4^{2-} (1)
		Molecular $CuSO_4$ (1)
		e-transfer (1)
LU 3: Moderate understanding (4)	Inaccurate (3)	Deposition of SO_4^{2-} (1)
	Just described (1)	e-transfer (2)
		Molecular $CuSO_4$ (1)
LU 2: Weak understanding (7)	Inaccurate (7)	Deposition of SO_4^{2-} (7)
		No precipitation of Cu (4)
		No ionization of Zn (3)
		e-transfer (2)
		Molecular $CuSO_4$ (1)
LU 1: Alternative understanding (10)	Inaccurate (3)	Deposition of SO_4^{2-} (4)
	Accurate (2)	No precipitation of Cu (3)
	Just described (5)	No ionization of Zn (2)
		Molecular $CuSO_4$ (2)
		e-transfer (4)

words, even if they talked about what the animation is depicting, they did not indicate whether it was an accurate animation or not. In general, they mostly referred to submicroscopic level features that cannot be observed; thus, these were not directly related to the experimental evidence. For example, a participant, who showed a weak understanding of redox reactions, as conveyed by the storyboard representation, shown in Fig. 2.3, just described the animation by focusing on the submicroscopic level features and stated in the interview that:

> Electron transition between zinc and copper ions. Precipitation of copper one-to-one ratio between copper and leaving zinc ions…. (LU 2—just described)

It was also observed that preservice teachers who only described the animation either pointed out that the animation was not showing some specific information about the redox reaction they viewed in the experimental video, or they conveyed their alternative conceptions. For the former case, for instance, one of the preservice teachers described the animation by pointing out what the animation is not showing, stating that:

> It does not explain how electrical conductivities changed. (LU 3—just described)

Another preservice teacher who only described the animation, demonstrated an alternative conception about the characteristics of macroscopic and submicroscopic features, as well as making connection between them. In fact, this participant and others expected to see some kind of color change in the animation, that is to say the decrease in the number of copper ions at the submicroscopic level did not mean color change to them, as represented in the following quote:

> In the experiment, color change occurs in the solution, we cannot see this difference in the animation. (LU 3—just described)

It was very surprising that there was only one preservice teacher with a moderate understanding described the animation without identifying its accuracy. This particular preservice teacher seemed to have difficulty in making connection between macroscopic and submicroscopic features, because the colors presented in the submicroscopic level animation appeared her to be analogous to the colors of the physical entities shown in the experimental video. She stated in the interview that:

> It made me thought that, [in the animation], the color of zinc [atoms] gets darker, so what is depositing must be zinc. (LU 3—just described)

Interestingly, half (50%) of the preservice teachers with an alternative understanding of redox reactions were able to identify the animation as accurate without necessarily identifying any experimental evidence depicting in the animation. It might be claimed that even though they had some alternative conceptions about the redox reactions in the beginning, they were able to identify how the reaction should proceed after viewing the animation. In other words, viewing and critiquing this animation helped them refine their understanding. The following quotes represent the views of two different preservice teachers with an alternative understanding, but indicated that the animation was accurate:

> It is more complicated and more effective animation. It's really helpful in all aspects. (LU 1—accurate)
>
> How a redox reaction takes place was a bit more shaped in my mind. So far we learned the phenomenon mathematically focusing on the charge changes in ions; for example, it changes from (2+) to (0), or it changes from (0) to (2+). I have never thought how these entities move at the molecular level. I can say that I am kind of enlightened. (LU 1—accurate)

After Viewing the Inaccurate Animation

Similarly, the analysis of reflective writings and interviews about the inaccurate animation revealed that 14 of 22 preservice teachers identified this animation as *inaccurate*, two of the preservice teachers categorized it as *accurate*, and six of them *just described* and discussed what they saw in the animation without categorizing it as accurate or inaccurate. The number of participants who categorized the scientifically inaccurate animation as accurate, inaccurate or just described at each level of understanding, and the features they highlighted are shown in Table 2.3.

The results of the analysis showed that there were two preservice teachers with an alternative understanding considered the inaccurate animation as accurate. The storyboards generated by these preservice teachers can be seen in Fig. 2.4. One of these preservice teachers with an alternative understanding did not even refer to redox reaction in her initial storyboard, and the other one described the redox reaction, mostly at the symbolic level (Fig. 2.4). Both seemed to have lack of understanding of redox reactions, and that might be the reason why they found the inaccurate animation to be accurate. The following quote comes from the interview of one of these preservice teachers:

> This [inaccurate animation] makes more sense. Because sulfate accompanies copper (+2) and there, it takes electrons from zinc, and zinc becomes (+2) charged, oxidized. Then it [SO_4^{-2}] remains on the metal. Very good, then copper takes its electrons and goes back to the solution. (LU 1—accurate).

The majority of the preservice teachers referred to submicroscopic features when they argue that the animation was inaccurate. Among these features, deposition of sulfate ions was the mostly referred one (13), electron transfer (9), no precipitation of copper (7), no ionization of zinc (5), and molecular form of $CuSO_4$ (5). The following quotes from interviews represent how preservice teachers express these features while categorizing the animation as inaccurate:

> This animation seemed to be conceptually wrong to me, because it shows copper and sulphate together, as if they were bonded, as if they were not ionized in the solution. (LU 4—inaccurate)

> I think this animation is wrong because Cu(0) didn't give precipitation and stayed as ion [surrounded] with [by] water molecules. (LU 2—inaccurate)

> I think this [animation] includes a misconception. …. None of the zinc [atoms] move to the solution. (LU 2—inaccurate)

Some of the preservice teachers noticed these submicroscopic features, but they did not indicate if the animation was accurate or inaccurate, instead, they just described. This might be because they were not sure about how the redox reactions were taking place. The following quote from the reflective writing of a preservice teacher who just described the animation:

> The SO_4 ions are aggregating on the zinc metal. How the electron transfer is happening is shown in between the Zinc and SO_4 ions. (LU 3—just described).

In general, the preservice teachers critiqued both animations, and they were mostly able to identify the accurate and inaccurate animations by referring to the features justified or refuted by experimental evidence. The categorizations of the accurate and inaccurate animations with respect to the preservice teachers' levels of understanding are summarized in Table 2.4.

As shown in Table 2.4, even though in the beginning the preservice teachers did not have a scientific understanding regarding redox reactions, as they critiqued and reflected on the animations, they were mostly able to identify the accurate and inaccurate animations. Specifically, considering the preservice teachers with a moderate understanding, only one preservice teacher could not categorize the animations. Among the preservice teachers with a weak understanding, all of them were able to identify the scientifically inaccurate animation as inaccurate, the majority (71%)

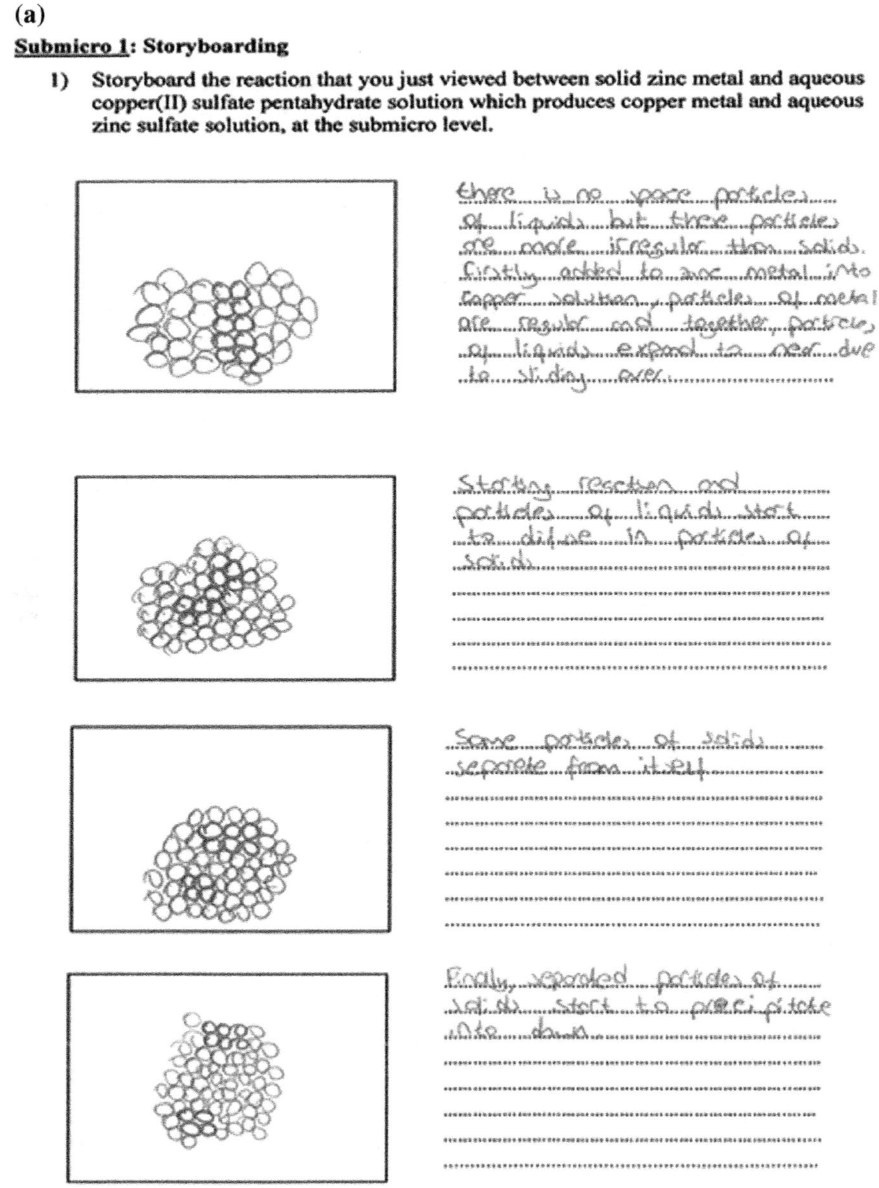

(a)

Submicro 1: Storyboarding

1) **Storyboard the reaction that you just viewed between solid zinc metal and aqueous copper(II) sulfate pentahydrate solution which produces copper metal and aqueous zinc sulfate solution, at the submicro level.**

there is no space particles of liquids but these particles are more irregular than solids. firstly added to zinc metal into copper solution, particles of metal are regular and together, particles of liquids exposed to near due to sliding over

Starting reaction and particles of liquids start to difuse in particles of solids

Some particles of solids separate from itself

Finally separated particles of solids start to precipitate into down

Fig. 2.4 Storyboard of a preservice teacher with an alternative understanding of redox reaction

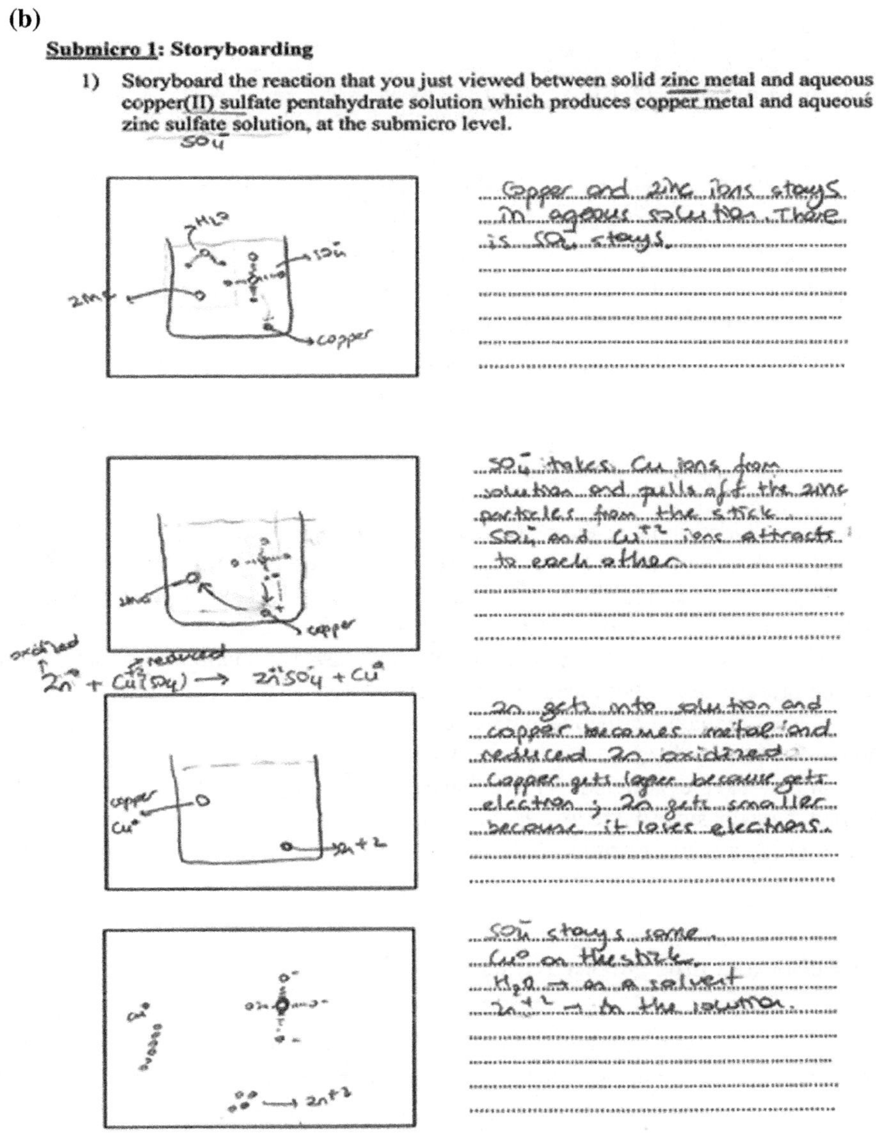

Fig. 2.4 (continued)

Table 2.4 Summary of preservice teachers' categorization of the accurate and inaccurate animations according to their level of understanding

Level of understanding, LU (N)	Categorization of the scientifically *accurate* animation (N)	Categorization of the scientifically *inaccurate* animation (N)
LU 4: Scientific understanding (1)	Accurate (1)	Inaccurate (1)
LU 3: Moderate understanding (4)	Accurate (3)	Inaccurate (3)
	Just described (1)	Just described (1)
LU 2: Weak understanding (7)	Accurate (5)	Inaccurate (7)
	Just described (2)	
LU 1: Alternative understanding (10)	Accurate (5)	Inaccurate (3)
	Inaccurate (1)	Accurate (2)
	Just described (4)	Just described (5)

was able to identify the scientifically accurate animation, and only two of them just described the accurate one. For the preservice teachers with an alternative understanding, half of them (50%) recognized the scientifically accurate animation, some of them (40%) were just described, and only one of them considered this animation as scientifically inaccurate. The scientifically inaccurate animation was much harder for the preservice teachers with an alternative understanding, because few of them (30%) were recognized the inaccurate features, half of them (50%) just described, and two of them believed it as accurate. Thus, it can be claimed that, while critiquing the animations, the participants might have passed through a mental dissonance that helped them consider the redox reactions more in depth and revisit their understanding resulting a more coherent level of understanding.

Conclusions and Implications

This study aimed to investigate how preservice chemistry teachers critique and respond to two dynamic submicroscopic animations of a redox reaction in which one shows a scientifically *accurate* process consistent with experimental evidence and another one shows scientifically *inaccurate* process of the same reaction. In the beginning, based on the preservice teachers' storyboard representations of a redox reaction, their understandings concerning redox reactions were identified in a 4-level scale from an alternative understanding, to a weak, a moderate, and a scientific understanding. It was observed that the majority of preservice teachers held a weak or an alternative understanding of redox reactions. This finding was found to be similar to the previous studies, because redox reactions have been difficult for students to understand as reflected in the frequent appearance of alternative conceptions in their representations of redox reactions (Acar & Tarhan, 2007; Brandriet & Bretz, 2014; Garnett & Treagust, 1992; Österlund & Ekborg, 2009; Rosenthal & Sanger, 2012;

Sanger & Greenbowe, 1997; Schmidt & Volke, 2003). Moreover, given the average GPA of the participants (2.60 out of 4.00), it might be suggested that they did not have a strong chemistry background. Thus, this can be a limitation of this study.

This conceptual difficulty of redox reactions often makes chemistry educators utilize various tools and methods to help students visualize them, especially at the submicroscopic level. Dynamic visualizations, namely animations, were mostly used in teaching redox reactions revealing improved learning outcomes (Kelly, 2014; Kelly et al., 2017; Lowe, 2003; Rosenthal & Sanger, 2012). Yet, allowing students only to view such visuals may not always result in substantial learning or change in their understandings (Akaygun & Jones, 2013; Kelly, 2014; Tasker & Dalton, 2006; Rosenthal & Sanger, 2012, 2013). In the current study, not only the participants viewed conceptually conflicting dynamic submicroscopic animations of a redox reaction but also critiqued and reflected on these animations by responding to the scaffolding questions provided to them. In this respect, when the participants viewed the scientifically accurate animation, the participant with a scientific understanding (one of 22) was able to connect the macroscopic features of a redox reaction with the submicroscopic occurrences as well as identifying the scientific features in the scientifically accurate animation (see Table 2.2 and Table 2.3). About two-thirds of the participants with a moderate, a weak and an alternative understanding (13 of 21) identified this animation as accurate, but the participants with a weak and an alternative understanding particularly exhibited difficulties associating the macroscopic features of a reaction with the submicroscopic occurrences. For example, few of these participants pointed the paling color of copper(II) sulfate solution as an experimental evidence, and they rarely connected this evidence with the decreasing number of copper ions in the solution. Perhaps, these participants recognized the scientifically accurate animation as accurate by noticing its particular scientific features. However, this does not necessarily imply that these participants with low levels of initial understandings completely made sense of and adopted all features of scientifically accurate animation into their conceptual framework (Vosniadou, 1994). They likely dissatisfied with their own understandings (Posner et al., 1982) and started to create new understandings of redox reactions based on their current experience (Bussey et al., 2013). Moreover, about one-third of the participants with a moderate, a weak and an alternative understanding (seven of 21) did not clearly indicate their views about the scientifically accurate animation, but they only described its particular features without referring to its accuracy. These participants probably were not willing to give up their own incomplete or nonscientific ideas and then hold onto the scientific one (Posner et al., 1982). In other words, they likely noticed some reasonable features in the scientifically accurate animation (Bussey et al., 2013), but they might have been unsure if the reaction process happens as it was represented in the animation. There might also exist some features in the scientifically accurate animation, conflicting with their own understandings of redox reactions.

When the participants viewed the scientifically inaccurate animation, the participant with a scientific understanding (one of 22) was able to identify this one as inaccurate, and she pointed out its nonscientific features by offering certain reasons. About two-third of the participants with a moderate, a weak and an alternative understanding (13 of 21) identified scientifically inaccurate animation as inaccurate, and the majority of these participants were able to identify the nonscientific features of the animation and justify how such features should have been accurately represented. This indicated that even if these participants initially held incomplete or nonscientific understandings of redox reactions, with comparing and contrasting the certain features of two animations, in other words with critically viewing the two animations, they were able to distinguish the scientific one from nonscientific one. This actually revealed that these participants possibly dissatisfied with their own initial understandings and started developing more scientific ones based on their experience with conceptually conflicting animations (Bussey et al., 2013; Posner et al., 1982; Vosniadou, 1994). However, two of the 10 participants with an alternative understanding found scientifically inaccurate animation as accurate, and half of the participants with an alternative understanding (five of 10) were just described the certain features of the scientifically inaccurate animation instead of clearly labelling the animation as accurate or inaccurate. Among the participants with an alternative understanding who just described the features of inaccurate animation probably viewed some of its features consistent with their own understandings but at the same time found some other features confusing. Thus, they likely held onto their dominant nonscientific ideas, but due to some confusing features of the animation, they avoided naming the animation as accurate or inaccurate.

It has been suggested that engaging students in critiquing the animations may encourage them to become mentally more active and may result in substantial conceptual progress (Chang & Linn, 2013). This might have happened as students experience cognitive dissonance and probably pass through conceptual change process (Bussey et al., 2013; Posner et al. 1982; Treagust & Duit, 2008). This is consistent with findings from the current study such that critiquing conceptually conflicting submicroscopic animations helped preservice teachers revise or refine their understandings. In other words, critically comparing and contrasting conceptually conflicting animations with guiding questions likely scaffolded the preservice teachers' learning as well as promoting their reflection on animations. The critically viewing dynamic submicroscopic animations as well as reconsidering the existing understandings with reflection can be suggested as an effective way of using animations for learning (Quintana et al., 2004).

Chemistry instruction predominantly emphasizes symbolic level, and consequently, students often memorize chemical equations without understanding that such an equation stands for a chemical reaction involving bond breaking and formation as well as particle rearrangement (Kelly & Hansen, 2017). However, students need to make connections between macroscopic and submicrocopic levels, so that they were better able to conceptualize chemical phenomenon. In this study, preservice teachers were asked to make connection between an experimental observation and submicroscopic level animations as they sought evidence for the redox reaction. That was not very easy for them, thus some participants only referred to the submicroscopic features without making connection or were able to identify only some evidence, not all of them. This might have happened because they were not very used to make such connections, instead they might have learnt the concepts at different levels in a more isolated manner. Thus, it is suggested to include macroscopic-submicroscopic level connections while teaching chemical concepts.

Acknowledgements This study was co-funded by The Scientific and Technological Research Council of Turkey (TUBITAK), Grant Number: 115K025, and National Science Foundation (NSF), Grant Number: 1525557. We would like to thank our collaborators Prof. Resa Kelly, Dr. Sarah Hansen and Dr. Adrian Villalta-Cerdas, and the project assistant Gözde Yıldırım, for their contribution and inspiration.

References

Acar, B., & Tarhan, L. (2007). Effect of cooperative learning strategies on students' understanding of concepts in electrochemistry. *International Journal of Science and Mathematics Education, 5,* 349–373.

Ainsworth, S. E. (1999). The functions of multiple representations. *Computers & Education, 33,* 131–152.

Akaygun, S., Adadan, E., & Kelly, R. (2019). Capturing preservice chemistry teachers' visual representations of redox reactions through storyboards. *Israel Journal of Chemistry*. https://doi.org/10.1002/ijch.201800133.

Akaygun, S., & Jones, L. L. (2013). Research-based design and development of a simulation of liquid–vapor equilibrium. *Chemistry Education Research and Practice, 14*(3), 324–344.

Ardac, D., & Akaygun, S. (2004). Effectiveness of multimedia-based instruction that emphasizes representations on students' understanding of chemical change. *Journal of Research in Science Teaching, 41,* 317–337.

Atkins, P., & Jones, L. L. (2010). *Chemical principles: The quest for insight* (5th ed.). New York: Freeman Company.

Brandriet, A. R., & Bretz, S. L. (2014). The development of the redox concept inventory as a measure of students' symbolic and particulate redox understandings and confidence. *Journal of Chemical Education, 91*(8), 1132–1144.

Bussey, T. J., Orgill, M., & Crippen, K. J. (2013). Variation theory: A theory of learning and a useful theoretical framework for chemical education research. *Chemistry Education Research and Practice, 14,* 9–22.

Chang, H.-Y., & Linn, M. C. (2013). Scaffolding learning from molecular visualizations. *Journal of Research in Science Teaching, 50*(7), 858–886.

Cheng, M., & Gilbert, J. K. (2009). Towards a better utilization of diagram in research into the use of representative levels in chemical education. In J. K. Gilbert & D. F. Treagust (Eds.), *Multiple representations in chemical education* (pp. 55–73). Dordrecht: Springer.

de Jong, O., Acampo, J., & Verdonk, A. (1995). Problems in teaching the topic of redox reactions: Actions and conceptions of chemistry teachers. *Journal of Research in Science Teaching, 32,* 1097–1110.

de Jong, O., & Treagust, D. (2002). The teaching and learning of electrochemistry. In J. K. Gilbert, O. De Jong, R. Justi, D. F. Treagust, & J. H. van Driel (Eds.), *Chemical Education: Towards research-based practice* (pp. 317–337). Dordrecht: Kluwer.

Driver, R., Squires, A., & Wood-Robinson, V. (1994). *Making sense of secondary science: Research into children's ideas*. London: Routledge.

Garnett, P. J., & Treagust, D. F. (1992). Conceptual difficulties experienced by senior high school students of electrochemistry: Electric circuits and oxidation-reduction equations. *Journal of Research in Science Teaching, 29,* 121–142.

Glaser, B. (1965). The constant comparative method of qualitative analysis. *Social Problems, 12*(4), 436–445.

Hsieh, H.-F., & Shannon, S. E. (2005). Three approaches to qualitative content analysis. *Qualitative Health Research, 15*(9), 1277–1288.
Johnstone, A. H. (1991). Why is science difficult to learn? Things are seldom what they seem. *Journal of Computer Assisted learning, 7,* 75–83.
Kelly, R. M. (2014). Using variation theory with metacognitive monitoring to develop insights into how students learn from molecular visualizations. *Journal of Chemical Education, 91*(8), 1152–1161.
Kelly, R. M. (2017). Learning from contrasting molecular animations with a metacognitive monitoring activity. *Educación Química, 28,* 181–194.
Kelly, R. M., & Hansen, S. J. R. (2017). Exploring the design and use of molecular animations that conflict for understanding chemical reactions. *Química Nova, 40*(4), 476–481.
Kelly, R. M., Akaygun, S., Hansen, S. J. R., & Villalta-Cerdas, A. (2017). The effect that comparing molecular animations of varying accuracy has on students' submicroscopic explanations. *Chemistry Education Research and Practice, 18*(4), 582–600.
Kozma, R. (2003). The material features of multiple representations and their cognitive and social affordances for science understanding. *Learning and Instruction, 13*(2), 205–226.
Lowe, R. (2003). Animation and learning: Selective processing of information in dynamic graphics. *Learning and Instruction, 13,* 157–176.
Orgill, M. (2012). Variation theory. In N. Seel (Ed.), *Encyclopedia of the sciences of learning* (pp. 2608–2611). New York, NY: Springer.
Österlund, L. L., & Ekborg, M. (2009). Students' understanding of redox reactions in three situations. *Nordic Studies in Science Education, 5*(2), 115–127.
Plass, J. L., Homer, B. D., & Hayward, E. O. (2009). Design factors for educationally effective animations and simulations. *Journal of Computing in Higher Education, 21*(1), 31–61.
Posner, G. J., Strike, K. A., Hewson, P. W., & Gertzog, W. A. (1982). Accommodation of a scientific conception: Toward a theory of conceptual change. *Science Education, 66*(2), 211–227.
Quintana, C., Reiser, B. J., Davis, E. A., Krajcik, J., Fretz, E., Duncan, R. G., … Soloway, E. (2004). A scaffolding design framework for software to support science inquiry. *Journal of the Learning Sciences, 13*(3), 337–386.
Rosenthal, D. P., & Sanger, M. J. (2012). Student misinterpretations and misconceptions based on their explanations of two computer animations of varying complexity depicting the same oxidation-reduction reaction. *Chemistry Education Research and Practice, 13,* 471–483.
Rosenthal, D. P., & Sanger, M. J. (2013). How does viewing one computer animation affect students' interpretations of another animation depicting the same oxidation–reduction reaction? *Chemistry Education Research and Practice, 14*(3), 286–296.
Sanger, M. J., & Greenbowe, T. J. (1997). Common student misconceptions in electrochemistry: Galvanic, electrolytic, and concentration cells. *Journal of Research in Science Teaching, 34*(4), 377–398.
Schmidt, H. J., & Volke, D. (2003). Shift of meaning and students' alternative concepts. *International Journal of Science Education, 25,* 1409–1424.
Seufert, T. (2003). Supporting coherence formation in learning from multiple representations. *Learning and Instruction, 13,* 227–237.
Smith, J., diSessa, A., & Roschelle, J. (1993). Misconceptions reconceived: A constructivist analysis of knowledge in transition. *Journal of the Learning Sciences, 3,* 115–163.
Stevens, S. Y., Delgado, C., & Krajcik, J. S. (2010). Developing a hypothetical multi-dimensional learning progression for the nature of matter. *Journal of Research in Science Teaching, 47*(6), 687–715.
Stieff, M. (2011). Improving representational competence using molecular simulations embedded in inquiry activities. *Journal of Research in Science Teaching, 48*(10), 1137–1158.
Taber, K. S. (2013). Revisiting the chemistry triplet: Drawing upon the nature of chemical knowledge and the psychology of learning to inform chemistry education. *Chemistry Education Research and Practice, 14*(2), 156–168.

Tasker, R., & Dalton, R. (2006). Research into practice: Visualisation of the molecular world using animations. *Chemistry Education Research and Practice, 7*(2), 141–159.

Tasker, R., & Dalton, R. (2008). Visualizing the molecular world—design, evaluation, and use of animations. In J. K. Gilbert. M., Reiner, and M. Nakhleh (Eds.), *Visualization: Theory and practice in science education* (pp. 103–131). Dordrecht: Springer.

Treagust, D. F., & Duit, R. (2008). Conceptual change: A discussion of theoretical, methodological and practical challenges for science education. *Cultural Studies of Science Education, 3*(2), 297–328.

Vosniadou, S. (1994). Capturing and modeling the process of conceptual change. *Learning and Instruction, 4*(1), 45–69.

Williamson, V. M., & Abraham, M. R. (1995). The effects of computer animation on the particulate mental models of college chemistry students. *Journal of Research in Science Teaching, 32*(5), 521–534.

Wu, H.-K., Krajcik, J. S., & Soloway, E. (2001). Promoting conceptual understanding of chemical representations: Students' use of a visualization tool in the classroom. *Journal of Research in Science Teaching, 38,* 821–842.

Zhang, Z. H., & Linn, M. C. (2011). Can generating representations enhance learning with dynamic visualizations? *Journal of Research in Science Teaching, 48,* 1177–1198.

Chapter 3
Reframing Chemistry Learning: The Use of Student-Generated Contexts

Edwehna Elinore S. Paderna, Rosanelia T. Yangco and Marlene B. Ferido

Abstract This chapter presents a teaching approach that focuses on student-generated contexts and how its use can reframe both chemistry teaching and learning. The eight phases of the Student-Generated Contexts Teaching Approach (SCTA) model—Introduction, Context Generation, Decision, Implementation, Presentation, Discussion, Reflection, and Context Regeneration—are described in terms of individual and collaborative context generation. To determine the nature of the contexts the students themselves generated, decision logs, reflection papers, and audio-recording transcripts were content-analyzed. The emerging themes on the nature of the contexts are: (1) source of the context; (2) level of the context; and (3) student engagement in the context. Sixty-five Grade 10 chemistry students comprised the two heterogeneous intact classes: one group was exposed to Individual Student-Generated Contexts Teaching Approach and another group was exposed to Collaborative Student-Generated Contexts Teaching Approach. The topics covered in the study included acids and bases, neutralization reactions, reduction–oxidation reactions, electrochemical cells, and electroplating. In this chapter, the contexts generated by the two groups are compared in terms of fluency, flexibility, and complexity. Finally, the benefits of using student-generated contexts in the teaching and learning of chemistry are discussed.

E. E. S. Paderna (✉) · R. T. Yangco
College of Education, University of the Philippines, Diliman, Philippines
e-mail: espaderna@up.edu.ph

R. T. Yangco
e-mail: rtyangco@up.edu.ph

M. B. Ferido
National Institute for Science and Mathematics Education Development,
University of the Philippines, Diliman, Philippines
e-mail: mbferido@up.edu.ph

M. Schultz et al. (eds.), *Research and Practice in Chemistry Education*,
https://doi.org/10.1007/978-981-13-6998-8_3

Introduction

Achieving scientific literacy is one of the main goals of science education (National Academy of Sciences, 2017). However, despite the call for scientific literacy, science assessment results provide alarming evidence that most students are not being prepared for the challenges ahead. Furthermore, students are learning very little science (Nelson, 2001, as cited in Child Trends Data Bank, 2013). They are taught to memorize facts and vocabulary and solve problems but almost never to connect the knowledge into a coherent picture of the physical world. Hence, it is necessary that chemistry education addresses this concern by fostering chemical literacy, with the ultimate goal that students find chemistry relevant to their daily lives.

Chemistry curricula in different countries like the United States of America, United Kingdom, Germany, Russia, Spain, Sweden, and Australia (Hill & Holman, 2000; Holman & Pilling, 2004; King, 2007; Nentwig, Demuth, Parchmann, Grasel, & Ralle, 2007; Sutman & Bruce, 1992) have highlighted the importance of making chemistry relevant to students. Similarly, in the Philippines, the K to 12 curriculum for Chemistry emphasizes that students should be able to demonstrate the understanding of the science concepts in chemistry, develop the skills to know the concepts, and apply the knowledge and skills to daily life (Department of Education, 2013).

In the light of the foregoing condition, an understanding of chemical literacy is essential. Such an understanding may be drawn from the domains of chemical literacy, namely: (1) scientific and chemical content knowledge; (2) chemistry in context; (3) higher-order learning skills; and (4) affective aspects (Shwartz, Ben-Zvi, & Hofstein, 2006). Chemistry in context is an important domain of chemical literacy in relation to this study. Context-based teaching in science has been explored in a number of studies (Nentwig et al., 2007; Palmer, 1997; Ramsden, 1997; Rayner, 2005; Roth & Roychoudhury, 1993), and results are favorable since students are more motivated to learn about chemistry. In addition, students find the subject relevant and feel that learning is more meaningful.

However, in the literature, contexts are usually given by the teacher to the students (Holman & Pilling, 2004; Nentwig et al., 2007; Palmer, 1997; Roth & Roychoudhury, 1993). What if the contexts are generated by the students themselves? Hence, this study developed the Student-Generated Contexts Teaching Approach which is founded on the constructivist learning theory and on context-based approach. It is a teaching approach wherein students construct their own contexts by referring to their related life experiences and to whatever they already know about the chemistry topic. The authors aimed to examine the contexts generated by the students. We then sought to determine the emerging themes that describe the nature of student-generated contexts.

Methodology

The participants in this study were 65 Grade 10 students enrolled in the University of the Philippines Integrated School (UPIS), which is the laboratory school of the University of the Philippines College of Education. The study was conducted during the third grading period (January to March 2017) of academic year 2016–2017. UPIS has only three Grade 10 sections. Two heterogeneous intact classes were randomly assigned to the teaching approaches—Individual Student-Generated Contexts Teaching Approach (ISCTA) and Collaborative Student-Generated Contexts Teaching Approach (CSCTA).

One of the authors was the lead researcher and she made arrangements with the two chemistry teachers handling the classes. The third grading syllabus and the schedule of the intervention period were discussed. Before the implementation of the intervention, the lead researcher observed the three classes to establish rapport with the students. During the observation, she noted that tardiness of students in the 7:00 a.m. class was a problem. As such, this class was purposively not included for random assignment to either of the ISCTA and CSCTA. This was done because the first 10–15 min of the chemistry period included the context generation. If some students in the 7:00 a.m. class were late, the context generation phase would have been greatly affected. In this regard, the two other classes were then randomly assigned to ISCTA and CSCTA.

The researchers followed the prescribed ethics guidelines of the University of the Philippines. A letter to the parents of students in the ISCTA and CSCTA classes was sent out. The letter informed the parents about the study and requested for their consent that their minor children be audiotaped and photographed during chemistry classes. Only students whose parents gave consent were audiotaped and photographed during the conduct of the study.

Moreover, a pre-intervention orientation on the Student-Generated Contexts Teaching Approach (SCTA) model was conducted to the ISCTA and CSCTA groups. The orientation was done to familiarize the students with the new teaching approach. The instructional phase, wherein the SCTA was implemented, lasted for 6 weeks. The two chemistry teachers of the school alternatingly observed the two classes to ensure that the teaching approaches were implemented as designed. After the intervention period, focus group discussions with the 12 groups of the ISCTA and CSCTA classes were held.

Transcripts from audiotaped discussions during context generation and focus group discussions, as well as student inputs from the decision logs and reflection papers, were content-analyzed and organized according to themes. Moreover, to analyze the nature of student-generated contexts, the contexts were evaluated in terms of: (1) fluency, which refers to the number of contexts generated; (2) flexibility, which refers to the number of different categories of contexts generated; and (3) complexity, which refers to the demand for application of more concepts/principles (Liu & Yu, 2004).

Student-Generated Contexts Teaching Approach (SCTA) Model

The Student-Generated Contexts Teaching Approach (SCTA) is founded on the constructivist learning theory which espouses the idea that learning takes place when individuals construct their own knowledge and meanings from experiences and reflect about what is presented (Gagnon & Collay, 2006; Hsieh & Cifuentes, 2003; Pittman, 1999; Spier-Dance, Mayer-Smith, Dance, & Khan, 2005). It is also anchored on the context-based approach/learning (Belt, Leisvik, Hyde, & Overton, 2005; Gagnon & Collay, 2006; Nentwig et al., 2007) wherein contexts are used in the teaching and learning process. However, unlike in previous studies where contexts were given to students, the student-participants in this study either individually (ISCTA) or collaboratively (CSCTA) generated their own contexts.

The SCTA model (Figs. 3.1 and 3.2) was developed in this study. It consists of eight phases, namely: Introduction, Context Generation, Decision, Implementation, Presentation, Discussion, Reflection, and Context Regeneration. The first four phases are shown in Fig. 3.1, while the remaining four phases are shown in Fig. 3.2.

To facilitate students' generation of contexts, the topic, learning objectives, and guide questions are presented in the introduction phase. The students then access their related life experiences to generate contexts during the context generation phase. A student exposed to ISCTA writes down his or her own generated contexts. On the other hand, students exposed to CSCTA generate contexts through sharing and discussion, which are audio-recorded and written on the decision log sheet.

The ISCTA and CSCTA groups mainly differed in the Context Generation and Context Regeneration phases. In the ISCTA class, the two phases were done individually by students, while in the CSCTA class, students were grouped to perform the two context generation activities. In both ISCTA and CSCTA, groups designed their own context-based activities during the Decision phase and implemented the planned activity the following day. It is important to note that the ISCTA and CSCTA groups did not do the ready-made recipe-type laboratory activities usually done in science classes in UPIS. However, the learning objectives in each of the five cycles in the SCTA were drawn from the objectives of the laboratory activities prepared by the chemistry teachers.

Results and Discussion

To determine the nature of the contexts generated by the students exposed to Individual Student-Generated Contexts Teaching Approach (ISCTA) and Collaborative Student-Generated Contexts Teaching Approach (CSCTA), audio transcripts of the context generation and context regeneration, decision logs, and reflection papers were content-analyzed. The contexts generated are summarized in Table 3.1. It is important to note that the contexts were generated prior to any input by the teacher, and

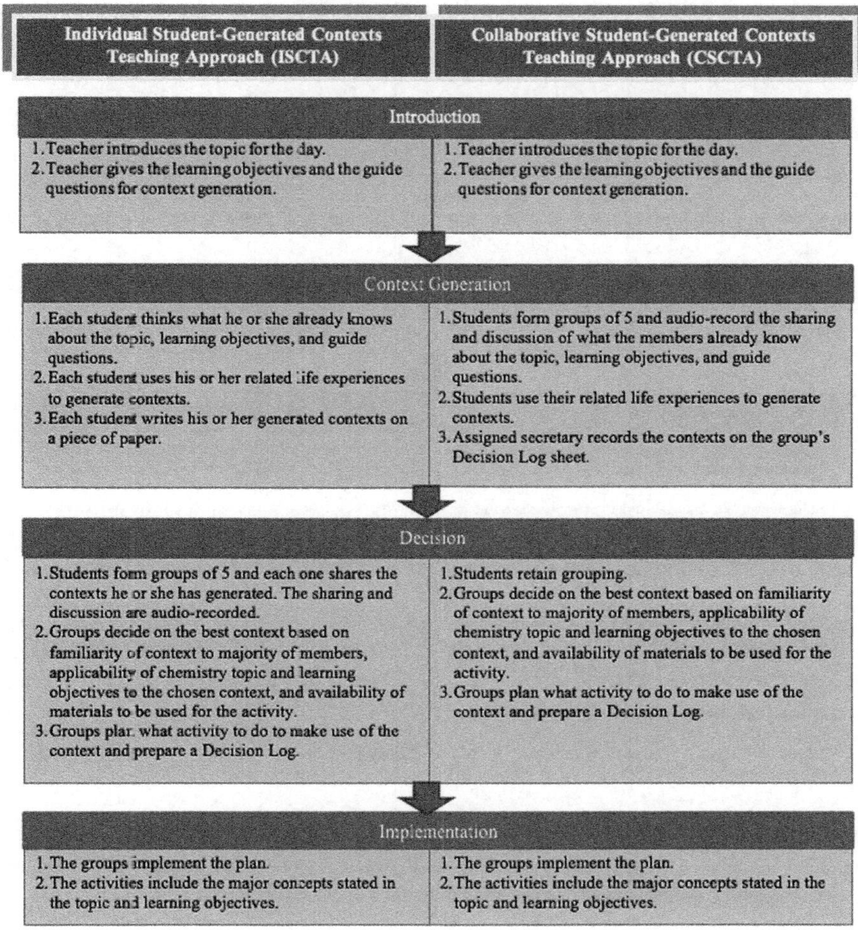

Fig. 3.1 Introduction, context generation, decision, and implementation phases of the student-generated contexts teaching approach model (Individual vs. Collaborative)

so there are contexts in the students' output that are wrong or inaccurate, indicating students' misconceptions. These answers from the students are marked.

The contexts generated by students exposed to ISCTA and CSCTA were examined in terms of fluency (the number of contexts generated), flexibility (the number of different categories of contexts generated), and complexity (demanding application of more concepts/principles for solving) (Liu & Yu, 2004). It was observed that, in general, the ISCTA class generated more contexts (fluency), which were more complex (complexity) than those generated by the CSCTA class. For example, in the topic Electrochemical Cells , the ISCTA class was able to give contexts related

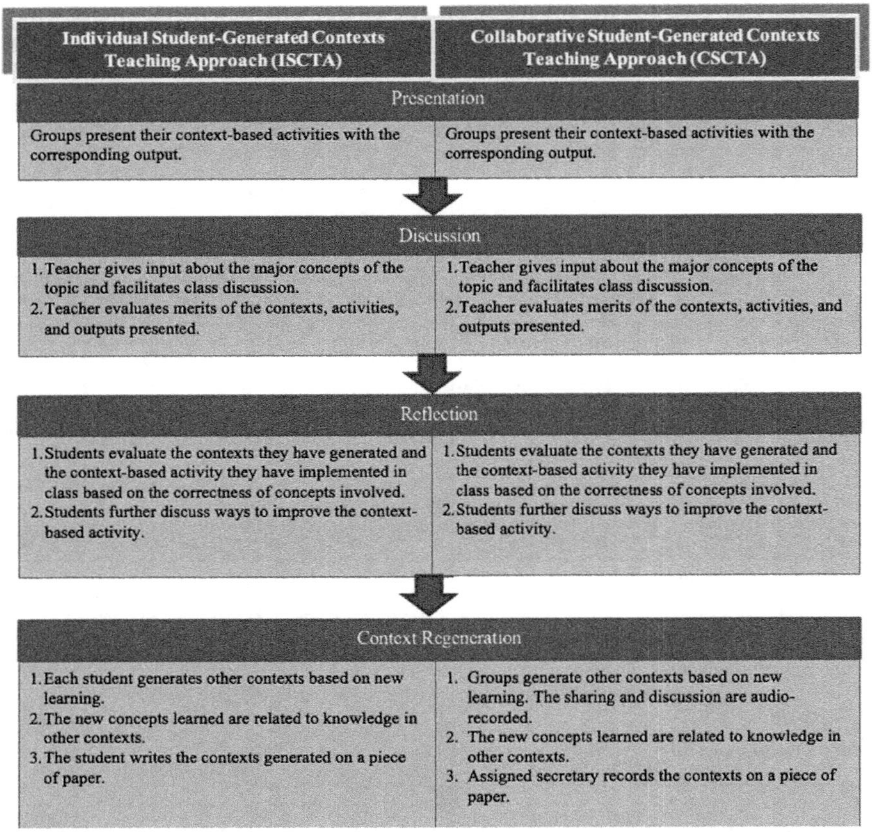

Fig. 3.2 Presentation, discussion, reflection, and context regeneration phases of the student-generated contexts teaching approach model (Individual vs. Collaborative)

to energy transformations and flow of electrons in the battery. However, in terms of the number of categories of contexts (flexibility), the ISCTA and the CSCTA groups were relatively the same.

Why did students exposed to ISCTA generate greater number of and more complex contexts? Students in the ISCTA class generated contexts individually and therefore were more focused on the task. This result is consistent with that of Baruah and Paulus (2008) who found that individual idea generation before a group brainstorming can be beneficial in terms of the number of ideas generated. This is supported by student feedback about the individual context generation, such as:

> It's better to generate the contexts individually because we were able to concentrate more in recalling our experiences about the topic. –ISCTA Student 5
>
> It's better individually because there's no distraction. –ISCTA Student 13

Table 3.1 Comparison of contexts generated by students in the ISCTA and CSCTA groups

Topic	ISCTA	CSCTA
Acids and Bases		
In what situations have you experienced acids? bases?	Washing, cleaning, eating, drinking, taking medicine	Cleaning, taking medicine, drinking, cooking, washing dishes
How do you describe acids? bases?	Vinegar, detergent, muriatic acid, hand soap, antacid, milk, nail polish,[a] polish remover,[a] lemon, feminine wash, vitamin C, calamansi	Vinegar, bleach, soft drinks, vitamin C, laundry detergent, baking soda, dishwashing liquid, antacid, muriatic acid, soap, lemon, calamansi, milk, stomach acid
	Litmus paper, pH > 7, pH < 7, corrosive, sour, base neutralizer, acid neutralizer, litmus paper turns blue, bitter, slippery	Litmus paper, pH scale
Neutralization Reactions		
In what situations have you experienced neutralization reaction?	Hyperacidity, antacid counters acidity, putting cold water in hot water,[a] putting water in lemon juice,[a] adding water to soap,[a] killing bacteria[a]	Soft drinks and Mentos,[a] taking antacid for hyperacidity, mixing water with fruit juice, bleach, dishwashing soap,[a] gardening (soil), hydrogen peroxide on wound,[a] antacid lowers acidity
Based on your experiences, what happens in a neutralization reaction?	A process where an acid and base become neutral (pH 7), acid and base react with each other, volcano models with vinegar and baking soda, salt as product of neutralization, equal amount of acid and base,[a] carbon dioxide is released[a]	Vinegar and baking soda science experiment
Reduction-Oxidation Reactions		
In what situations have you experienced reduction-oxidation reactions?	Rust, browning of apple and potato, painting metals, tarnishing of metallic jewelry/accessories	Browning of apple, pear, banana when exposed to air, cut eggplant right before cooking so it won't darken, rusting of metals, evaporation of mothballs[a]

(continued)

Table 3.1 (continued)

Topic	ISCTA	CSCTA
Based on your experiences, what happens in a reduction-oxidation reaction?	Metals rust because of oxidation, a chemical reaction, combustion, transfer of electrons, corrosion	Combustion, burning, photosynthesis
Electrochemical Cells (Batteries)		
1. In what situations have you used batteries?	Positive and negative terminals, batteries leak, overheat and explode, non-rechargeable and rechargeable batteries, different types of batteries, batteries used for gadgets, flashlights, remote controls, clocks and toys, battery runs out, source of power	Batteries used in phones, clocks, toys, remote controls, flashlights and cameras, batteries leak, overheat, swell and explode, two terminals, rechargeable batteries
2. What are your experiences with batteries?	Chemical to electrical energy, chemical to mechanical energy, chemical energy inside the battery, electrons move inside the battery, electron flows, battery stores energy	Electrons moving, science experiment, cathode and anode, related to electrolytes, fruit batteries
3. Based on your experiences, how do you think batteries work?		
Electroplating		
What are your experiences with electroplated (e.g., gold plated, silver plated, copper plated, etc.) materials?	Jewelry, silverware, coating, electroplated cars, accessories, medals, knives, electroplated jewelry darkens in time	Coins, electroplating jewelry and cars, galvanized nails and sheets, silverware, used in manufacturing different goods, electroplated jewelry tarnishes in time
Based on your experiences, how do you think electroplating (e.g., gold plating, silver plating, copper plating, etc.) works?	Metal deposited on the surface, prevents corrosion, alters conductivity, uses electricity, making the metal ions bond with other metal[a]	Decorating or coating metals to prevent corrosion

[a]Students' misconceptions

More interesting is that when asked during an FGD session, some students from the CSCTA felt that generating contexts individually would have been better for them.

> I feel it's individual because one will be forced to think of contexts. If by group, sometimes others just keep quiet and don't suggest, and expect that only others will speak. –CSCTA Student 19
>
> For me, individual also because you can think better since it's just you, just yourself. It's you who think. It's difficult if you join others because what you think can change.) –CSCTA Student 4

Moreover, it was observed that some students were playful and distracted. Some students were just listening to the others and not giving input to the group discussion. Accordingly, students in the CSCTA group remarked during an FGD session that:

> Although we are able to share while generating the contexts, we produce only few. Not all gives their own[sic]. Some just copy contexts and not think. –CSCTA Student 23
>
> Some are pressured, and some are shy. –CSCTA Student 15
>
> Some are lazy. –CSCTA Student 3

With these observations and feedback, the CSCTA may be improved by implementing stricter rules during the context generation. Moreover, collaborative context generation should be monitored more closely. The proposed modifications are intended for individual students, though generating contexts as a group, to be more focused and productive.

Nature of Student-Generated Contexts

Based on the contexts generated by the students in the five learning cycles, there are three emerging themes, namely: (1) source of context; (2) level of context; and (3) student engagement in the context. The themes answer the questions: (1) Where does the student's context come from? (2) How intimate and familiar is the student to the context? and (3) What is the student's participation in the context? The three emerging themes when combined present the interaction between the student and his or her contexts and describe the nature of student-generated contexts. Figure 3.3 summarizes the nature of student-generated contexts.

Source of Student-Generated Context

In the Context Generation phase the students were instructed to refer to their experiences that relate to the topic. Students were also given guide questions that facilitated their context generation. Given the said premise, the sources of the contexts generated by the students were identified as coming from their experiences at home, in school,

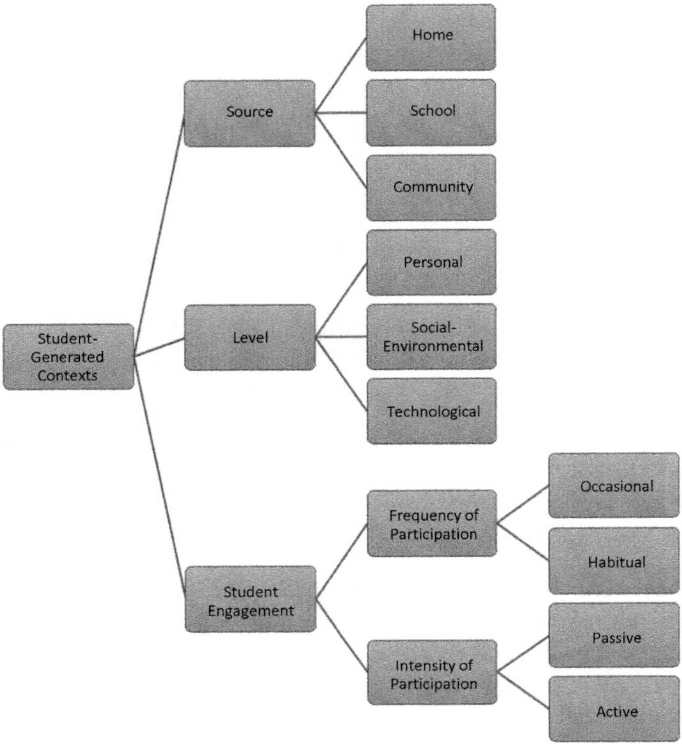

Fig. 3.3 The nature of student-generated contexts

and in the community. For example, students' contexts on acids and bases included common household items like vinegar, soap, and milk; and on reduction-oxidation reactions included apple, potato, and eggplant.

On the other hand, contexts that were sourced from school are those that they have experienced in previous grade level science classes. For instance, students mentioned contexts that pertain to experiments done in class like the use of litmus paper and the volcano model with vinegar and baking soda. Students also gave science or chemistry terms, which included pH scale, photosynthesis, combustion, chemical energy, among others, that are primarily learned in the science classroom.

Furthermore, contexts that came from the students' community were also mentioned. Example of community contexts included electroplated cars and galvanized nails and sheets. Table 3.2 summarizes the student-generated contexts of the ISCTA and CSCTA groups as classified according to source.

It is noticeable that most of students' contexts come from their experiences at home, followed by their experiences in school, and lastly in their community. Since context refers to "students' everyday experience of the physical world" (Szeto, 2008, p. 1), it is understandable that one's everyday experience begins and ends at home. Thus, students' generated contexts are mostly from their homes. Moreover, the con-

Table 3.2 Sources of student-generated contexts of ISCTA and CSCTA groups

Topic	Home		School		Community	
	ISCTA	CSCTA	ISCTA	CSCTA	ISCTA	CSCTA
Acids and Bases	Vinegar, detergent, muriatic acid, hand soap, antacid, milk, lemon, feminine wash, vitamin C, calamansi, washing, cleaning, eating, drinking, taking medicine	Vinegar, bleach, soft drinks, vitamin C, laundry detergent, baking soda, dishwashing liquid, antacid, muriatic acid, soap, lemon, calamansi, milk, stomach acid, cleaning, taking medicine, drinking, cooking, washing dishes	Litmus paper, pH > 7, pH < 7, corrosive, sour, base neutralizer, acid neutralizer, litmus paper turns blue, bitter, slippery	Litmus paper, pH scale		
Neutralization	Hyperacidity, antacid counters acidity	Taking antacid for hyperacidity, antacid lowers acidity, gardening (soil)	A process where an acid and base become neutral (pH 7), acid and base react with each other, volcano models with vinegar and baking soda, salt as product of neutralization	Vinegar and baking soda science experiment		
Reduction-Oxidation Reactions	Rust, browning of apple and potato, painting metals, tarnishing of metallic jewelry/accessories	Browning of apple, pear, banana when exposed to air, cut eggplant right before cooking so it won't darken, rusting of metals	Metals rust because of oxidation, a chemical reaction, combustion, transfer of electrons, corrosion	Combustion, burning, photosynthesis		

(continued)

Table 3.2 (continued)

Topic	Home		School		Community	
	ISCTA	CSCTA	ISCTA	CSCTA	ISCTA	CSCTA
Electrochemical Cells	Positive and negative terminals, batteries leak, overheat and explode, non-rechargeable and rechargeable batteries, different types of batteries, batteries used for gadgets, flashlights, remote controls, clocks, and toys battery runs out, source of power	Batteries used in phones, clocks, toys, remote controls, flashlights and cameras, batteries leak, overheat, swell, and explode, two terminals, rechargeable batteries	Chemical to electrical energy, chemical to mechanical energy, chemical energy inside the battery, electrons move inside the battery, electron flows, battery stores energy	Electron moving, science experiment, cathode and anode, related to electrolytes		
Electroplating	Jewelry, silverware, coating, electroplated accessories, medals, and knives, jewelry darkens in time	Coins, electroplating jewelry, silverware, electroplated jewelry tarnishes in time	Metal deposited on the surface, prevents corrosion, alters conductivity, uses electricity	Decorating or coating metals to prevent corrosion	Electroplated cars	Galvanized nails and sheets, used in manufacturing different goods

texts that come from students' experiences in school are "formalized and academic" experiences, which are covered by academic lessons and formal schooling.

The student-generated contexts in terms of the first emergent theme, *source of student-generated contexts*, show that, in general, students exposed to ISCTA generated greater number and more complex contexts derived from experiences at home, in school, and in community. However, it is notable that the CSCTA group was able to generate more contexts sourced from home for the topics acids and bases, neutralization, and reduction–oxidation reactions. This suggests that the group generation of contexts was more productive for the said topics. In terms of the number of context categories, the two groups are comparable.

Level of Student-Generated Context

The preceding analysis leads to the second theme, which is the level of context. Examining the contexts generated by the students gives three different levels: (1) personal; (2) social-environmental; and (3) technological. The three levels are consistent with the contexts identified by Holman and Pilling (2004). However, they identified social and environmental contexts separately, whereas in this study, the two are identified as one level. The student-generated contexts are classified into the three levels in Table 3.3.

At the personal level, contexts are those that the students themselves use like household items or observe like browning of fruits. In other words, these are the contexts that are closest or most familiar to the students. At the social-environmental level, on the other hand, contexts generated are those that are primarily shared with other people or contexts that are experienced with the environment. Examples of contexts at the social-environmental level are the science experiments done by the students with their classmates and the photosynthesis occurring in plants. Lastly, at the technological level, the contexts generated are those that involve technology such as decorating or coating metals and producing galvanized nails and sheets.

Based on the preceding classification into levels, it is evident that the student-generated contexts are mostly at the personal and technological levels. Since the personal level is the closest and most familiar to students, majority of the contexts are classified at this level. In addition, the results suggest that the student-participants in this study also have relatively rich experiences in technology. This result is not surprising since students of this generation are reasonably well-versed in terms of technology, which in this study refers to both products (e.g., batteries, electroplated cars, galvanized nails and sheets) and processes (e.g., neutralization, electrolysis, electroplating).

Comparing the ISCTA and CSCTA groups and similar to earlier results, the fluency and complexity of contexts are in favor of the ISCTA group. Furthermore, the two groups are comparable in terms of flexibility of contexts generated.

Table 3.3 Levels of student-generated contexts of ISCTA and CSCTA groups

Topic	Personal		Social-Environmental		Technological	
	ISCTA	CSCTA	ISCTA	CSCTA	ISCTA	CSCTA
Acids and Bases	Vinegar, detergent, muriatic acid, hand soap, antacid, milk, lemon, feminine wash, vitamin C, calamansi, sour, bitter, slippery, washing, cleaning, eating, drinking, taking medicine	Vinegar, bleach, soft drinks, vitamin C, laundry detergent, baking soda, dishwashing liquid, stomach acid, cleaning, taking medicine, drinking, cooking, washing dishes	Corrosive		Litmus paper, pH > 7, pH < 7, pH scale, base neutralizer, acid neutralizer	Litmus paper, pH scale
Neutralization	Hyperacidity, antacid counters acidity	Taking antacid for hyperacidity, antacid lowers acidity	Volcano models with vinegar and baking soda	Gardening (soil), vinegar and baking soda science experiment	Salt as product of neutralization	Neutralization
Reduction-Oxidation Reactions	Rust, browning of apple and potato, painting metals, tarnishing of metallic jewelry/accessories	Browning of apple, pear, banana when exposed to air, cut eggplant right before cooking so it won't darken, rusting of metals	Metals rust because of oxidation, a chemical reaction, combustion, corrosion	Combustion, burning, photosynthesis		Reduction-Oxidation Reactions

(continued)

Table 3.3 (continued)

Topic	Personal		Social-Environmental		Technological	
	ISCTA	CSCTA	ISCTA	CSCTA	ISCTA	CSCTA
Electrochemical Cells	Positive and negative terminals, batteries leak, overheat and explode, non-rechargeable and rechargeable batteries, different types of batteries, batteries used for gadgets, remote controls, clocks flashlights and toys, battery runs out, source of power	Batteries used in phones, clocks, toys, remote controls, flashlights and cameras, batteries leak, overheat, swell and explode, two terminals, rechargeable batteries		Science experiment, fruit batteries	Chemical to electrical energy, chemical to mechanical energy, chemical energy inside the battery, electrons move inside the battery, electron flows, battery stores energy	Electrons moving, cathode and anode, related to electrolysis
Electroplating	Jewelry, silverware, coating, accessories, medals, knives, electroplated jewelry darkens in time	Coins, silverware, electroplated jewelry tarnishes in time			Electroplated cars, metal deposited on the surface, prevents corrosion, alters conductivity, uses electricity	Electroplating jewelry and cars, galvanized nails and sheets, used in manufacturing different goods, decorating or coating metals to prevent corrosion

Student Engagement in Student-Generated Context

Student engagement in context describes how a student is as a participant in a particular context. It is twofold, covering the frequency and intensity of participation. For instance, in terms of frequency, a student may be a habitual or an occasional participant in the context. On the other hand, in terms of intensity of participation, one may be either an active or a passive participant in the context. The contexts generated by the student-participants are categorized in Table 3.4.

Contexts that are included in habitual participation by students are in relation to everyday activities like the use of soap and dishwashing liquid and access to different types of batteries. In terms of intensity of participation, students are passive on contexts that involve chemical reactions and processes like combustion, corrosion, and photosynthesis. On the other hand, students are active participants in using household items like in washing, cleaning, and eating.

In comparison, the ISCTA group generally generated greater number and more complex contexts in both levels of student engagement than the CSCTA group. However, students exposed to CSCTA were more of active participants in the topic neutralization and electrochemical cells. As in the themes sources and levels of contexts, the two groups generated the same number of context categories.

Conclusions and Implications for Practice

Since students exposed to Individual Student-Generated Contexts Teaching Approach (ISCTA) were able to generate greater number and more complex contexts compared to students exposed to Collaborative Student-Generated Contexts Teaching Approach (CSCTA), teachers can think of ways on how to maximize the context generation in CSCTA. One suggestion is for teachers to efficiently monitor the student generation of contexts because students tend to be distracted, playful, shy, and lazy at times during the task. Team roles could also be used to make sure that all members of the group are performing a certain assigned task.

Moreover, the sharing of contexts during context generation gives opportunity for students to explain and make others explain as well as link different units of information. Teachers can see to it that students are sharing and discussing, explaining his or her contexts and making others explain their contexts. In this way cognitive elaboration and associating skills will be enhanced.

The student-generated contexts that have been gathered in this study can be used by teachers in preparing activities in class since the said contexts are the ones to which students are able to find meanings and so, are likely to be more interesting for them. Similarly, teacher trainers for chemistry can use the contexts in preparing materials for trainings. Curriculum developers can incorporate student-generated contexts in the chemistry curriculum to emphasize concepts based on what students find most relatable and relevant to them. In the same way, textbook writers can adopt a

Table 3.4 Sources of student-generated contexts of ISCTA and CSCTA groups

Topic	Frequency of Participation				Intensity of Participation			
	Occasional		Habitual		Passive		Active	
	ISCTA	CSCTA	ISCTA	CSCTA	ISCTA	CSCTA	ISCTA	CSCTA
Acids and Bases	Muriatic acid, bleach, antacid, calamansi	Muriatic acid, bleach, antacid, calamansi, lemon, baking soda	Vinegar, detergent, hand soap, milk, lemon, feminine wash, vitamin C, laundry detergent, dishwashing liquid	Vinegar, soft drinks, vitamin C, laundry detergent, dishwashing liquid, soap, milk, stomach acid	Litmus paper, pH > 7, pH < 7, corrosive, base neutralizer, acid neutralizer, litmus paper turns blue	Litmus paper, pH scale	Washing, cleaning, eating, drinking, taking medicine, sour, bitter, slippery	Cleaning, taking medicine, drinking, cooking, washing dishes
Neutralization	Hyperacidity, antacid counters acidity	Taking antacid for hyperacidity, antacid lowers acidity			A process where an acid and base become neutral (pH 7), acid and base react with each other, salt as product of neutralization, antacid counters acidity	Antacid lowers acidity	Volcano models with vinegar and baking soda	Vinegar and baking soda science experiment, taking antacid for hyperacidity
Reduction-Oxidation Reactions	Rust, browning of apple and potato	Browning of apple, pear, banana when exposed to air			Metals rust because of oxidation, a chemical reaction, combustion, transfer of electrons, corrosion, browning of apple and potato	Combustion, burning, photosynthesis, browning of apple, pear, banana when exposed to air	Painting metals	Cut eggplant right before cooking so it won't darken

(continued)

Table 3.4 (continued)

Topic	Frequency of Participation				Intensity of Participation			
	Occasional		Habitual		Passive		Active	
Electrochemical Cells	Batteries leak, overheat and explode	Batteries leak, overheat, swell and explode, science experiment	Positive and negative terminals, non-rechargeable and rechargeable batteries, different types of batteries, batteries used for gadgets, flashlights, remote controls, clocks and toys, battery runs out, source of power	Batteries used in phones, clocks, toys, remote controls, flashlights and cameras, two terminals, rechargeable batteries	Chemical to electrical energy, chemical to mechanical energy, chemical energy inside the battery, electrons move inside the battery, electron flows, battery stores energy	Electrons moving, cathode and anode, related to electrolytes	Batteries used for gadgets, flashlights, remote controls, clocks and toys	Batteries used in phones, clocks, toys, remote controls, flashlights and cameras, science experiment, fruit batteries
Electroplating	Jewelry, silverware, coating, accessories, medals, knives, electroplated jewelry darkens in time	Silverware, electroplated jewelry tarnishes in time		Coins	Metal deposited on the surface, prevents corrosion, alters conductivity, uses electricity, jewelry darkens in time, electroplated cars	Decorating or coating metals to prevent corrosion, electroplating jewelry and cars, galvanized nails and sheets, used in manufacturing different goods, electroplated jewelry tarnishes in time	Jewelry, silverware, accessories, medals, knives	Silverware

framework that uses student-generated contexts as a starting point in the development of concepts. This will allow students to be drawn to the material more easily since they can relate to the context.

Acknowledgements Thank you very much to: Philippine Department of Science and Technology for the Ph.D. Scholarship; University of the Philippines Diliman Chancellor Michael L. Tan for the Dissertation Aid and the Research Dissemination Grant; University of the Philippines Diliman Office of the Vice-Chancellor for Research and Development through its Vice-Chancellor Fidel R. Nemenzo for the Dissertation Grant; and U.P. College of Education Scholarship Committee for the Dissertation Grant.

References

Baruah, J., & Paulus, P. B. (2008). Effects of training on idea generation in groups. *Small Group Research, 39*, 523–541.

Belt, S. T., Leisvik, M. J., Hyde, A. J., & Overton, T. L. (2005). Using a context-based approach to undergraduate chemistry teaching—a case study for introductory physical chemistry. *Chemistry Education Research and Practice, 6*(3), 166–179.

Child Trends Data Bank. (2013). Science proficiency: Indicators on children and youth. https://www.childtrends.org/wp-content/uploads/2012/10/10_Science_Proficiency.pdf. Acccessed December, 2018.

Department of Education. (2013). *K-12 curriculum (Science)*. Pasig City: Bureau of Secondary Education, Philippines Department of Education.

Gagnon, G. W., Jr., & Collay, M. (2006). *Constructivist learning design*. California: Corwin Press.

Hill, G., & Holman, J. (2000). *Chemistry in context* (5th ed.). Cheltenham: Nelson Thornes.

Holman, J., & Pilling, G. (2004). Thermodynamics in context: A case study of contextualized teaching for undergraduates. *Journal of Chemical Education, 81*(3), 373–375.

Hsieh, Y. J., & Cifuentes, L. (2003). A cross-cultural study of the effect of student-generated visualization on middle school students' science concept learning in Texas and Taiwan. *Educational Technology Research and Development, 51*(3), 90–95.

King, D. (2007). Teacher beliefs and constraints in implementing a context-based approach in chemistry. *Teaching Science, 53*(1), 14–18.

Liu, Y.-H., & Yu, F.-Y. (2004). Active learning through student generated questions in physics experimental classrooms. Paper presented at International Conference on Engineering Education, October 2004, Gainesville, Florida.

National Academy of Sciences. (2017). National science education standards: An overview. Retrieved from http://books.nap.edu/html/nses/htmloverview.html.

Nelson, G. D. (2001). Remarks on the release of the NAEP 2000 science assessment results. AAAS Project 2061. American Association for the Advancement of Science. Retrieved from http://www.project2061.org/newsinfo/press/rl011120a.htm.

Nentwig, P. M., Demuth, R., Parchmann, I., Grasel, C., & Ralle, B. (2007). Chemie im context: Situating learning in relevant contexts while systematically developing basic chemical concepts. *Journal of Chemical Education, 84*(9), 1439–1444.

Palmer, D. (1997). The effect of context on students' reasoning about forces. *International Journal of Science Education, 19*(6), 681–696.

Pittman, K. M. (1999). Student-generated analogies: Another way of knowing? *Journal of Research in Science Teaching, 36*(1), 1–22.

Ramsden, J. M. (1997). How does a context-based approach influence understanding of key chemical ideas at 16+? *International Journal of Science Education, 19*(6), 697–710.

Rayner, A. (2005). Reflections on context-based science teaching: A case study of physics for students of physiotherapy. In *Uniserve science blended learning symposium proceedings*. Retrieved from http://science.uniserve.edu.au/pubs/procs/wshop10/2005Rayner.pdf.

Roth, W. M., & Roychoudhury, A. (1993). The development of science process skills in authentic contexts. *Journal of Research in Science Teaching, 30*(2), 127–152.

Shwartz, Y., Ben-Zvi, R., & Hofstein, A. (2006). The use of scientific literacy taxonomy for assessing the development of chemical literacy among high school students. *Chemistry Education Research and Practice, 7*(4), 203–225.

Spier-Dance, L., Mayer-Smith, J., Dance, N., & Khan, S. (2005). The role of student-generated analogies in promoting conceptual understanding for undergraduate chemistry students. *Research in Science and Technological Education, 23*(2), 163–178.

Sutman, F. X., & Bruce, M. H. (1992). Chemistry in the community: A five-year evaluation. *Journal of Chemical Education, 69*(7), 564–567.

Szeto, A. K. (2008). Teaching science content, concepts and processes with contexts. In *Ideas on Teaching*. Centre for Development of Teaching and Learning, National University of Singapore. http://www.cdtl.nus.edu.sg/Ideas/iot100.htm Accessed December, 2018.

Chapter 4
Interactive Immersive Virtual Reality to Enhance Students' Visualisation of Complex Molecules

Mihye Won, Mauro Mocerino, Kok-Sing Tang, David F. Treagust and Roy Tasker

Abstract Computer-aided visualisation has played an essential role in helping students construct a better understanding of molecular interactions over the years. However, students still need to translate 2D images on a screen into 3D molecular structures in their minds, and they often find it challenging to develop a clear mental picture of complex molecules. As an alternative platform for molecular visualisation, immersive virtual reality (VR) holds great promise for addressing students' difficulties. To examine the educational possibilities and limitations of immersive VR environments, we adopted a design research approach. We developed a series of learning activities to teach students about an enzyme and its interactions with substrates and then conducted a user study in an immersive VR environment. Twenty-two first-year university chemistry students participated in the user study. Their VR sessions along with pre- and post-interviews were video-recorded and later analysed to understand how various features of the immersive VR activities supported and impeded students' conceptual understanding and their ability to visualise molecules. Future research directions and the implications for chemistry teaching are discussed.

M. Won (✉) · K.-S. Tang · D. F. Treagust
School of Education, Curtin University, Perth, Australia
e-mail: mihye.won@curtin.edu.au

K.-S. Tang
e-mail: kok-sing.tang@curtin.edu.au

D. F. Treagust
e-mail: d.treagust@curtin.edu.au

M. Mocerino
School of Molecular and Life Sciences, Curtin University, Perth, Australia
e-mail: m.mocerino@curtin.edu.au

R. Tasker
Department of Chemistry, College of Science, Purdue University, West Lafayette, IN, USA
e-mail: rtasker@purdue.edu

M. Schultz et al. (eds.), *Research and Practice in Chemistry Education*,
https://doi.org/10.1007/978-981-13-6998-8_4

Visualisation Challenges in Learning Chemistry

Visualisation of molecular structures and interactions is an integral part of learning chemistry, but students often find it challenging to develop a clear mental image of scientific phenomena affecting non-observable entities (Gilbert & Treagust, 2009; Treagust, Chittleborough, & Mamiala, 2003). When students look at the formulae or diagrams of molecules on a piece of paper, they may not be able to translate the 2D images into 3D molecular structures because visualisation involves a high level of visuospatial thinking skills and background knowledge (Wu & Shah, 2004).

To support students' visualisation of molecular structures and processes, chemistry educators have adopted various models and media and investigated their effects (Jones, 2013; Tasker, 2017). The most common visualisation tool would be the ball-and-stick models (e.g. Molymod®) where students can physically manipulate and connect atoms to build a molecule. While they are effective in showing 3D molecular structures, they do not represent the dynamic nature of chemical interactions between molecules and it is difficult to build complex molecules with them. Computer-based visualisation tools overcome those limitations: chemical formula visualisation tools (e.g. Avogadro or Chimera) allow users to build complex molecules and examine their structures; animation and simulation programs (e.g. Connected Chemistry or VisChem) allow students to manipulate different variables and observe the dynamic molecular processes. Yet, computer-based visualisation tools display 3D objects on a 2D computer screen and students still need to process 2D images to construct concrete 3D molecular structures. The limitations of conventional visualisation tools become more obvious when dealing with complex molecules, such as proteins. Figure 4.1 shows how an enzyme (acetylcholinesterase) and a substrate (acetylcholine) are positioned relative to each other. As readers can imagine, it is challenging to visualise the depth and scale of the complex molecules (e.g. enzymes) and their dynamic interactions with other molecules (e.g. substrates) from looking at those 2D images on a computer screen.

As an alternative platform to support visualisation of chemical structures and processes, immersive virtual reality (VR) shows great promise in overcoming the limitations of conventional media (Limniou, Roberts, & Papadopoulos, 2008). In immersive VR environments, virtual objects and their surroundings appear realistic and interactive to allow users to suspend their sense of physical surroundings and fully immerse themselves in their interactions with virtual objects (Biocca & Delaney, 1992). In such environments, learners can inspect and interact with 3D visualisations to learn abstract concepts that are not easily represented through conventional media (Dede, 2009). For chemistry visualisation, instead of moving a computer mouse to see different angles on a screen to figure out the 3D structures of molecules, in immersive VR environments users can grab and rotate molecules and turn their heads around to examine the molecular structures and interactions, as observed in Fig. 4.2.

Due to its superior visualisation capabilities, the 3D immersive visualisation environment has been used to explore potential protein–drug interactions for research chemists (Anderson & Weng, 1999; Tse, Li, Leung, Lee, & Wong, 2011), and also to

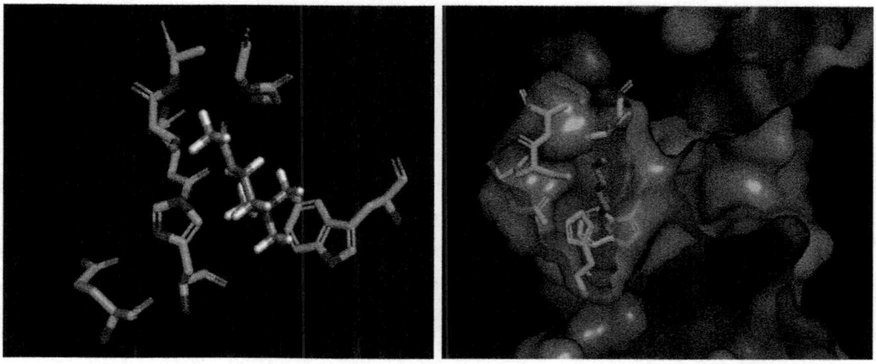

Fig. 4.1 The interaction between the substrate (acetylcholine in pink) and the active site (key amino acids in green) of acetylcholinesterase presented on a computer screen without (left image) and with (right image) the protein surface. Image reprinted with permission from the author

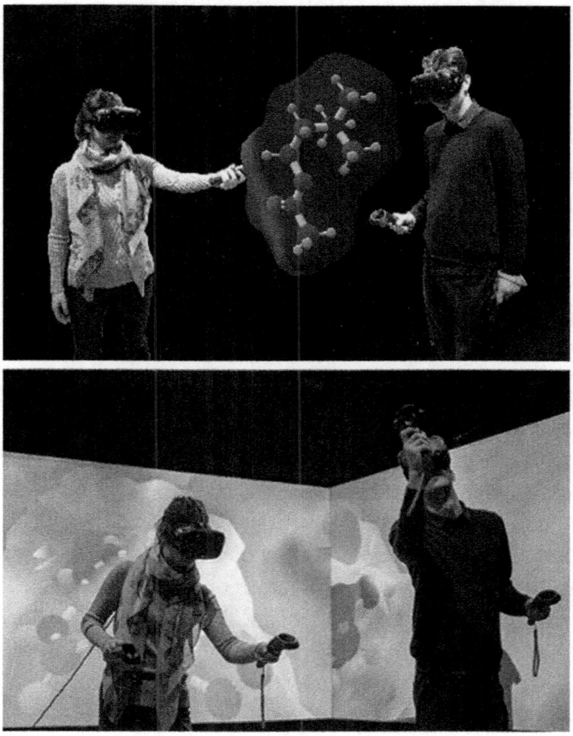

Fig. 4.2 Students exploring the molecular structures in an immersive VR environment. Image reprinted with permission from the author

investigate its educational possibilities for medical education (e.g. Hansen. 2008) and for industrial training (e.g. Gavish et al., 2013). However, immersive VR technology has attracted little attention from the general public, curriculum material developers, and instructors until recently (Dede, Jacobson, & Richards, 2017). With the advancement in high density displays and graphics processing units, the display resolution and functionality of the immersive VR environment has improved immensely in recent years to make 3D visualisation more realistic, interactive, and affordable. There have been no significant efforts to investigate the educational benefits of a 3D immersive VR environment for undergraduate students struggling to visualise and understand protein structures and interactions, making this research timely.

When investigating educational possibilities, researchers need to be mindful that a visualisation tool also poses challenges to learners (Won, Yoon, & Treagust, 2014). A new technological environment, in this case, an immersive 3D VR, could be too unfamiliar or cognitively overwhelming to its users to achieve the desired learning outcomes (Ainsworth, 2008; Ploetzner & Lowe, 2004). Especially when the topic is complex and new to the learner, dynamic visualisations in 3D may inadvertently cause or strengthen misunderstanding of relevant scientific concepts and phenomena (Lowe & Boucheix, 2017; Schwan & Papenmeier, 2017). Thus, it is important to explore both the possibilities and limitations of the 3D immersive environment in supporting students' visualisation and conceptual understanding. In doing so, this study aims to identify the beneficial features and the challenges associated with the use of molecular-level visualisations in immersive VR.

Design of Immersive VR Activities

In designing dynamic visualisation applications, researchers tend to rely on "the designers' intuitions and opinions" (Lowe & Boucheix, 2017, p. 6) rather than evidence-based design principles and guidelines (Scheiter, 2017). For more effective development of dynamic visualisation applications of molecular worlds, it is necessary to document and investigate various design features to encourage students to focus their attention on the key ideas and construct their scientific understanding (Jones & Kelly, 2015). Based on the literature on the design frameworks of immersive learning environments (Dede, 2009; Liu, Dede, Huang, & Richards, 2017), the authors of this paper identified key elements to include in the immersive VR activities to increase learning motivation and focus students' attention on key learning objectives.

Collaborative construction of knowledge: Based on the social constructivist learning theory, collaborative group work tasks were incorporated into the immersive VR application. These tasks encouraged students to move through the activities together and to discuss their ideas with their partners to come up with answers. Research into dynamic visualisation suggests that if learners are encouraged to collaborate and ask questions with more constructivist approaches, students' conceptual understanding and learning experiences improved considerably (McElhaney, Chang,

Chiu, & Linn, 2014). Immersive virtual reality research studies also suggest that when participants interact in a shared virtual environment to complete the tasks together, they feel more drawn to the learning activities through social immersion (Gardner & Sheaffer, 2017; Krämer, 2017).

Identification and alignment of the learning objectives with the activities: As discussed above, conventional media have limited visualisation capabilities for complex 3D materials such as enzymes. Because of these limitations, enzyme-substrate reactions are often taught with a simple lock-and-key model without providing a realistic understanding of how the enzyme works from a mechanistic point of view. An immersive 3D visualisation environment offers a unique learning opportunity for students to understand the complexity and scale of a protein and its reactions with realistic representations. As students explored the complex molecules and their reactions, the authors of this paper hoped that students would grasp one of the fundamental concepts of chemistry—that the structure and the electron density of molecules play major roles in chemical reactions—and included various prompts and questions to direct students' attention to these aspects.

Engaging and relevant context: The structure of acetylcholinesterase and its reaction with acetylcholine may appear too complex for the target audience (first-year chemistry students). However, the authors of this paper regard it as a strongly relevant context to engage students in an immersive VR environment due to its biological importance and its relation to contemporary world affairs. Acetylcholine is a neurotransmitter that is released by nerve cells to enable muscles to contract. Once the message has been received, acetylcholinesterase breaks down the neurotransmitter into its components, acetate and choline. As acetylcholinesterase plays a key role in human bodies, it has been targeted by snake venoms and chemical weapons. Sarin, a nerve agent recently used as a chemical weapon in the Middle East, binds irreversibly to acetylcholinesterase, disabling the enzyme and causing death in minutes.

Taking into consideration the features discussed above, the project team drew a series of drawings with narrations and instructions in a storyboard. The final storyboard included three main activities. The first activity was constructing the acetylcholine molecule and predicting/examining the electron density of the molecule (Fig. 4.3a). This activity served multiple purposes: (1) to help students get to know their partners; (2) become familiar with the interface of the VR application; and (3) to directly compare their experience of molecule construction within the immersive VR environment to their experience of the same activity with a physical ball-and-stick model (Molymod®). Students predict the electron-rich and electron-poor areas of the molecule and compare their answer with the electron density map (Fig. 4.3b).

For the second activity, students' progress in exploring the structure of the acetylcholinesterase enzyme by zooming in and out (Fig. 4.3c), rotating, and walking around the enzyme. They then compare different representations of the enzyme by toggling between them (e.g. surface model with electron density map, ball-and-stick model, carbon wireframe model, and ribbon model with alpha-helices and beta-sheets) (Fig. 4.3d). While in this activity, students are prompted to find the deep gorge of the protein leading to the active site.

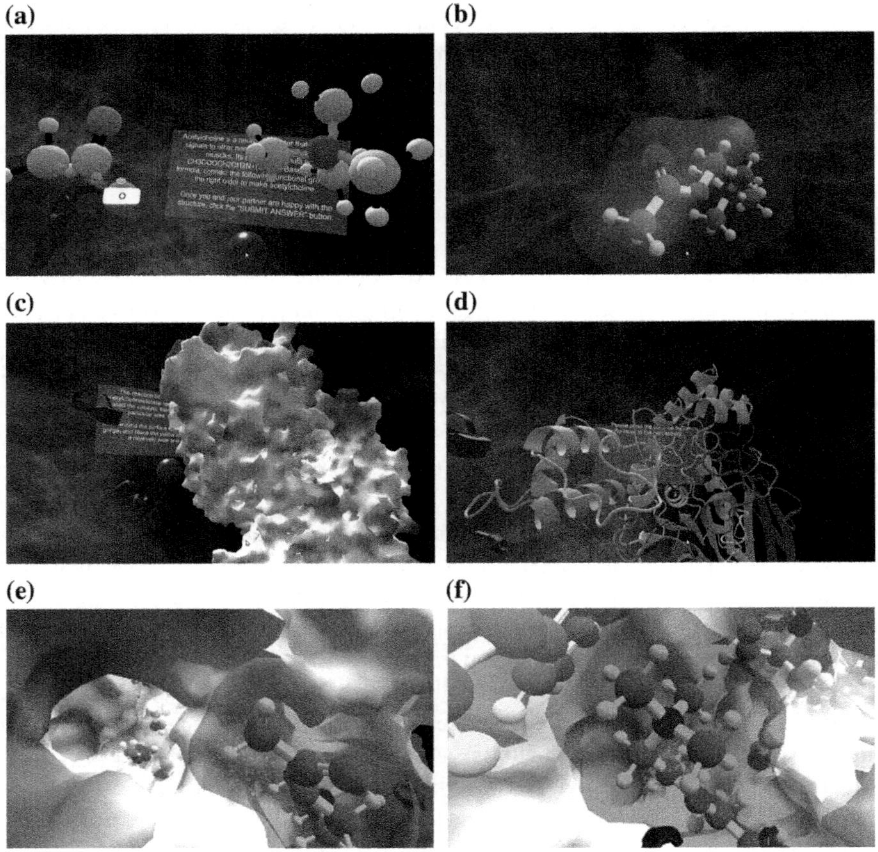

Fig. 4.3 Screenshots of the VR application. **a** construction of the acetylcholine molecule, **b** electron density map of the acetylcholine molecule, **c** exploration of the structure of the enzyme (acetylcholinesterase), **d** comparison of different representations of the enzyme, **e** placing acetylcholine at the entrance of the gorge, **f** placing acetylcholine at the active site of the enzyme. Image reprinted with permission from the author

In the third activity, they drag and place the acetylcholine molecule at the entrance of the gorge (Fig. 4.3e) and then walk through the cavity of the enzyme to place acetylcholine at the active site (Fig. 4.3f). For each task, they are prompted to consider the electron density and the structural fit between the enzyme and the substrate. Once the activities are done, they watch animations for the step-by-step reaction mechanisms in 3D.

Research Design

This study adopts a design research approach (Collins, Joseph, & Bielaczyc, 2004) to investigate how students are interacting within a 3D immersive VR environment for learning chemistry in relation to the affordances and design of the VR environment. Design research is particularly useful in the dual goals of advancing a general theory of how students learn from a VR immersive learning environment as well as improving the design of a specific VR application. Due to the reciprocal process of theory informing practice and vice versa, design research utilises an iterative cycle to design the VR application and understand students' learning experiences at the same time.

This paper reports the first iteration: the project team built a set of VR learning activities based on current theoretical knowledge of how students learn from a VR environment and first-year university students tried out the first version of the VR application. Qualitative exploratory data were collected and analysed to understand how students interacted within the VR environment and how they evaluated their learning experiences (Merriam & Tisdell, 2016; Treagust, Won, & Duit, 2014). Based on the outcome of the first cycle, the project will improve the learning tasks for the next iteration. Thus, with each iteration, we will gain more detailed knowledge of how the students interact with the VR environment as well as the affordances and design considerations of an immersive VR environment.

The project team designed a series of learning tasks with a storyboard, and an immersive VR application was developed using Unity®. Among various VR platforms, HTC VIVE™ was adopted in this study for its multi-user interactivity, accurate movement/position tracking, and a natural field of view. The ethics approval was sought and received from the research integrity office at an Australian university. Five beta-testing sessions were conducted, involving chemistry lecturers and third-year biochemistry students. Twenty-two first-year chemistry students (12 female and 10 male) volunteered to participate in the study. The students had a varying degree of background knowledge in chemistry (with/without high school chemistry). Several students mentioned they played video games regularly. Four students had tried simple VR headsets (e.g. Samsung Gear VR) at general electronics stores.

The user study sessions were comprised of three segments. First, students were given the chemical formula of acetylcholine, $CH_3COO(CH_2)_2N^+(CH_3)_3$, and told to construct the molecule using the Molymod® kit and predict the electron density of the molecule. The interviewer asked questions to assess students' background knowledge and their confidence in molecular structure and intermolecular forces. Students then wore the VR headsets and immersed themselves in the virtual molecular world. The participating students interacted with each other in a shared physical and virtual space. The virtual location was aligned with the physical space so that their interactions felt natural (instead of hearing disembodied voice). Please refer to Fig. 4.2 for the setup of the session. The immersive VR activities lasted for about an hour (40–75 min). After the VR activities, students were asked to evaluate their experience in terms of what they learnt and how the VR application could be improved. The post-interview lasted for 20 min for each pair. All user study sessions were video-recorded and observed by at least two researchers. In this paper, we discuss students' experiences with VR, based on researchers' observation and students' self-reports at the post-interview.

Results

Throughout the VR session, students were engaged in continuous conversations with their partners and immersed themselves in the 3D VR activities. As they took their VR headsets off, the common response was "It was amazing!"

Interviewer: How was [your experience with the VR]?

Students A&B: (together) It was really good fun.

Student B: It was so good to see what's happening. I'm a visual learner, so being able to see each step taking place was amazing.

Student A: It was nice to see the structure and to be able to move them, rotate them, and redo it. So, it was really nice being able to move it and have the pseudo tangibility of something that is normally in the domain of textbooks and YouTube.

They were most impressed by the immersive visualisation technology as it allowed them to explore the protein structure from different vantage points effortlessly and watch the reaction mechanisms in 3D. Students felt fully immersed in the virtual world (*sensory immersion*). They excitedly walked over and bent their bodies to find a better angle to look at particular sites of the protein or the reaction mechanism. As they walked through the gorge of acetylcholinesterase, students ducked their heads to avoid a collision with the virtual electron clouds over their heads. When students pressed hard on the zoom-in buttons to be completely surrounded by the grey electron clouds or the multitude of atoms, some of the students even exclaimed that they felt claustrophobic.

Interviewer: Did you feel it was real?

Student C: Yes, it did feel real. When I accidently zoomed in too fast, I thought the [electron] cloud would hit my face, which was why I was like 'whoa' (jerking her head back), moving back. (laughing)

Student D: (laughing together) and I forgot I was in a room, after a while.

In the following, we present students' responses in relation to the key design features.

Students' Collaborative Interactions Within the VR Environment

Immersive virtual reality inherently supports learning by doing rather than learning by observing only (Slater, 2017). Many participating students mentioned that they enjoyed the action-based tasks the most—building acetylcholine molecule like assembling Lego® blocks to see the electron density map, and walking inside the protein and dragging the acetylcholine molecule through the gorge to the catalytic site to see how they would react.

Student A: The most fun part for me was putting together the molecule. I really enjoyed that. […] If we had more of doing things [like that], it would have been more fun.

Interviewer: Before you did the VR, you basically did the same activity with the ball-and-stick model. […] So how was the activity different or similar in the VR environment?

Student B: It was bigger and brighter. (laughs)

Student A: Novel, it was novel. And I think it was more of working as a team. I know we worked as a team on this part (pointing at the Molymod kit in front of him), but in the VR, it was like we were double-checking it, tugging things so it wouldn't fall apart… it was a lot of fun doing that, piecing it together.

The participating students explained why they particularly enjoyed those action-oriented activities. Even though they had done the same activity of constructing acetylcholine with a ball-and-stick model in their pre-interview, constructing a molecule in the immersive environment felt much 'cooler' and more fun. Several students mentioned that it was a surreal experience handling the molecule pieces as disembodied spirits in a nice background of galaxy; others said it was easier to handle with bigger atoms and labels; and many others said because they were able to see the electron density map afterwards, they liked the VR activity more.

Similar explanations were given for placing the acetylcholine molecule at the active site: they had never imagined that they could walk through a protein (with a substrate in their hands) so it was a novel experience; it was intriguing to feel acetylcholine bouncing back or stuck at the gorge entrance and not getting through; and they were able to see afterwards if the relative position and orientation of acetylcholine was correct to start the enzyme-substrate catalytic reaction.

Simple but strategic actions coupled with immediate visual feedback within a novel immersive virtual reality environment increased their enjoyment and engagement in the activities. Dede (2009) calls this *actional immersion,* "empowering the participant in an experience to initiate actions impossible in the real world that have novel, intriguing consequences" (p. 66). Even though this design component was not explicitly considered by the project team (especially for the construction of the acetylcholine molecule activity), the relevant features were intuitively included in the VR application, enhancing students' experiences.

The project team identified *social immersion* as one of the key design elements for an immersive VR environment (Dede et al., 2017), and all of the tasks were designed for collaborative group work.

Interviewer: Which part of the VR activities was most helpful for you to understand the interaction between acetylcholine and acetylcholinesterase?

Student C: I think everything worked out well, but I think that it was good that we got to try it out ourselves first and then they tell us if it was right or wrong and show us what it actually is.

Student D: Yeah, and then we had time to talk about why we were wrong.

Student C: and get to figure out for ourselves why we were wrong.

Students noted that they greatly enjoyed being able to talk with their partner about the activities, and get things done together in the virtual space. Even though their faces or bodies were not displayed in the virtual space, as shown in Fig. 4.3b, they were able to 'see' their partner as they could recognise where their partner was and what they were pointing at from the position of their headset, controllers and voice. When their partner was pointing out interesting aspects of the protein from a different angle, they walked over to their partner's side to compare what they saw with what their partner saw. They did not hesitate to share their ideas with each other to come up with an answer. They checked if their partner was happy with the answer and if there were any alternative explanations.

The students were recruited from first-year chemistry workshops where students worked together in small groups to solve conceptual questions. Consequently, most participants had known their partners for a while and they knew how to interact with them. Those participants who did not know their partners beforehand were more hesitant in the beginning, but after 5–10 min, their interactions became natural, except for one pair of students who rarely voiced their thoughts and opinions, could not decide on their answers, and did not know how to proceed. Their VR session dragged on and drained energy from the participants. The VR activities were designed with the assumption that students would be able to discuss and negotiate their ideas and complete the tasks collaboratively. However, when the participants had limited negotiation skills, the power of social immersion was remarkably weakened.

Development of Conceptual Understanding in the VR Environment

Most students commented that the immersive VR activities transformed the abstract to the concrete (Slater, 2017). Before participating in the VR activities, enzymes were understood as bio-catalysts that play important roles in our bodies, but students did not have any concrete idea of how big they would be and what shape they would assume.

Interviewer How about the protein structure? Was it what you imagined it to be?

Student A: It's never been like, oh it's a protein structure and it looks like this (gestures as if he was looking at a bigger structure in space). It is a protein structure, and it has a binding structure with a lot of ligands, [...] you write R group (gestures as if he's writing something) and you forget about it. I didn't know actually what it was and where it existed.

If an enzyme had any shape at all in their minds, it was the block shape often seen in illustrative diagrams of lock-and-key mechanisms. When students first saw the enzyme in immersive VR, they did not recognise what they were seeing at first. Only after they zoomed in and out of the enzyme and changed the representations from the surface model to the ball-and-stick model and vice versa did they realise the shape and size of the enzyme. They were amazed by the fact that out of the hundreds of amino acids in the enzyme, only three amino acids (which were hidden inside a deep gorge) were involved in the catalytic reaction with its substrate. The complexity of the catalytic reaction mechanisms between the enzyme and its substrate also surprised the students.

Even though the majority of the participating students had taken human biology and/or chemistry at high school in addition to the first-year university chemistry, they admitted that they had had no idea what a protein would look like and how complex the enzyme-substrate reaction could be before the VR session.

Interviewer: What do you think you learnt from participating in this activity?

Student C: When I learnt about [acetylcholinesterase] in human biology, I imagined a little dot (gestures with her hands) breaking up acetylcholine, instead of what we saw just then.

Student D: For the protein structure, what we learnt in class was misleading a little bit—because it relies on your own perceptions of what it could be rather than actually showing what it is like. [...] I didn't really know before—how many reactions were involved in the [enzyme] reaction until they showed it to us at the end, like how everything happens. How it moves into.

Student C: ah, the triad thing.

Student D: yeah, and how it moves out of it. I didn't know any of that. And how many molecules are involved.

Student C: yeah, I was the same. I pretty much imagined an enzyme would be like a dot, going in like this, and breaking it up. But it is so different.

Student D: it is a lot more...

Students C&D: (together) complicated, yeah.

Student D: a lot more is involved that I didn't know before, or consider.

In addition to demonstrating the complexity of protein structures and enzyme reactions, the project aimed to help students understand the impact of molecular structure and electron density in chemical reactions, including the catalytic reaction between enzymes and substrates. Based on the colour coding scheme of the electron density map, students were able to identify electron-rich areas (in red) and electron-poor areas (in blue). Considering the electron density, they moved the molecules so that desired reactions could occur.

However, the influence of the molecular structure was less obvious to students. Despite the written and audio instructions to consider both the electron density map and the structure of molecules, most students focused their attention exclusively on the electron density map. Only two groups seriously considered the shape and size of acetylcholine in relation to the gorge entrance, the passage way, and the catalytic chamber. They walked back and forth through the gorge and surveyed the area from various angles to identify the best location and direction for acetylcholine to come through. Three groups realised that they had to consider the structure of acetylcholine and the gorge area only after they saw acetylcholine was not coming through in the direction they wished.

The fact that students did not consider the 3D structural aspects of the enzyme is puzzling because one of the main strengths of immersive VR is providing opportunities for students to explore objects in 3D. Is it simply because students were not used to considering the spatial aspects of molecules or is it because the spatial aspects were not highlighted enough to gain students' attention during the activity? This question needs to be explored in further studies.

Discussion and Implications

As chemistry educators, we constantly think about better ways to show molecular interactions to students. Overcoming the limitations of conventional visualisation media, immersive VR presents great promise as an alternative visualisation medium to support students' chemistry learning, especially for 3D spatial recognition of complex molecules. In an effort to examine the educational possibilities of VR, the project team designed a series of learning tasks to teach students about an enzyme and its reactions, integrating several design features for immersive VR environments, such as social immersion, actional immersion, and sensory immersion (Dede et al., 2017). This user study with first-year university chemistry students demonstrated that an immersive VR learning environment could seamlessly transform abstract concepts of an enzyme and its catalytic reactions into something concrete and tangible to encourage students to actively integrate and construct knowledge.

For future research, we intend to identify and integrate more key chemistry concepts and skills into immersive VR environments, and reach out to larger audiences to systematically investigate how various features of immersive VR interact with the diversity in students' conceptual understanding, visuospatial thinking skills, and social skills. The research on visualisation and immersive virtual reality is moving towards investigating pedagogical affordances of the medium rather than focusing on its technical affordances (Slater, 2017). The project intends to move forward by coordinating with university chemistry instructors to determine productive ways to integrate immersive VR in chemistry teaching and learning.

Acknowledgements This project was supported by the Research and Innovation Support Program (RISP17-02) at the School of Education, Curtin University. Adam Mathewson and Jesse Helliwell contributed to the development of the VR application. Jianye Wei and Vergel Mirana contributed to running the user study sessions. Hyerin Park synchronised the video and audio files for analysis. Curtin HIVE (Highly Immersive Visualisation eResearch) provided technical support and hosted the user trial sessions for the project.

References

Ainsworth, S. (2008). How do animations influence learning? In D. Robinson & G. Schraw (Eds.), *Current perspectives on cognition, learning, and instruction: Recent innovations in educational technology that facilitate student learning* (pp. 37–67). Charlotte, NC: Information Age Publishing.

Anderson, A., & Weng, Z. (1999). VRDD: Applying virtual reality visualization to protein docking and design. *Journal of Molecular Graphics and Modelling, 17*(3–4), 180–186. https://doi.org/10.1016/s1093-3263(99)00029-7.

Biocca, F., & Delaney, B. (1992). Immersive virtual reality technology. In F. Biocca & M. R. Levy (Eds.), *Communication in the age of virtual reality* (pp. 57–124). Hillsdale, NJ: Lawrence Erlbaum Associates.

Collins, A., Joseph, D., & Bielaczyc, K. (2004). Design research: Theoretical and methodological issues. *Journal of the Learning Sciences, 13*(1), 15–42. https://doi.org/10.1207/s15327809jls1301_2.

Dede, C. (2009). Immersive interfaces for engagement and learning. *Science, 323*(5910), 66–69. https://doi.org/10.1126/science.1167311.

Dede, C., Jacobson, J., & Richards, J. (2017). Introduction: Virtual, augmented, and mixed realities in education. In D. Liu, C. Dede, R. Huang, & J. Richards (Eds.), Virtual, augmented, and mixed realities in education (pp. 1–16). Smart Computing and Intelligence), Singapore: Springer.

Gardner, M. R., & Sheaffer, W. W. (2017). Systems to support co-creative collaboration in mixed-reality environments. In D. Liu, C. Dede, R. Huang, & J. Richards (Eds.), Virtual, augmented, and mixed realities in education (pp. 157–178). Smart Computing and Intelligence), Singapore: Springer.

Gavish, N., Gutiérrez, T., Webel, S., Rodríguez, J., Peveri, M., Bockholt, U., et al. (2013). Evaluating virtual reality and augmented reality training for industrial maintenance and assembly tasks. *Interactive Learning Environments, 23*(6), 778–798. https://doi.org/10.1080/10494820.2013.815221.

Gilbert, J. K., & Treagust, D. F. (Eds.). (2009). *Multiple representations in chemical education (Models and Modeling in Science Education)*. Dordrecht, Netherlands: Springer.

Hansen, M. M. (2008). Versatile, immersive, creative and dynamic virtual 3-D healthcare learning environments: A review of the literature. *Journal of Medical Internet Research, 10*(3), e26. https://doi.org/10.2196/jmir.1051.

Jones, L. L. (2013). How multimedia-based learning and molecular visualization change the landscape of chemical education research. *Journal of Chemical Education, 90*(12), 1571–1576. https://doi.org/10.1021/ed4001206.

Jones, L. L., & Kelly, R. M. (2015). Visualization: The key to understanding chemistry concepts. In M. V. Orna (Ed.), Sputnik to smartphones: A half-century of chemistry education (pp. 121–140). American Chemical Society.

Krämer, N. C. (2017). The Immersive power of social interaction. In D. Liu, C. Dede, R. Huang, & J. Richards (Eds.), *Virtual, augmented, and mixed realities in education* (pp. 55–70). Singapore: Springer.

Limniou, M., Roberts, D., & Papadopoulos, N. (2008). Full immersive virtual environment CAVE in chemistry education. *Computers & Education, 51*(2), 584–593. https://doi.org/10.1016/j.compedu.2007.06.014.

Liu, D., Dede, C., Huang, R., & Richards, J. (Eds.). (2017). *Virtual, augmented, and mixed realities in education (Smart Computing and Intelligence)*. Singapore: Springer.

Lowe, R., & Boucheix, J.-M. (2017). A composition approach to design of educational animations. In R. Lowe & R. Ploetzner (Eds.), *Learning from dynamic visualization* (pp. 5–30). Cham, Switzerland: Springer.

McElhaney, K. W., Chang, H.-Y., Chiu, J. L., & Linn, M. C. (2014). Evidence for effective uses of dynamic visualisations in science curriculum materials. *Studies in Science Education, 51*(1), 49–85. https://doi.org/10.1080/03057267.2014.984506.

Merriam, S. B., & Tisdell, E. J. (2016). *Qualitative research: A guide to design and implementation* (4th ed.). San Francisco: Jossey-Bass.

Ploetzner, R., & Lowe, R. (2004). Dynamic visualisations and learning. *Learning and Instruction, 14*(3), 235–240. https://doi.org/10.1016/j.learninstruc.2004.06.001.

Scheiter, K. (2017). Design of effective dynamic visualizations: A struggle between the beauty and the beast? Commentary on parts I and II. In R. Lowe & R. Ploetzner (Eds.), *Learning from dynamic visualization* (pp. 233–251). Cham, Switzerland: Springer.

Schwan, S., & Papenmeier, F. (2017). Learning from animations: From 2D to 3D? In R. Lowe & R. Ploetzner (Eds.), *Learning from dynamic visualisation*. Cham, Switzerland: Springer.

Slater, M. (2017). Implicit learning through embodiment in immersive virtual reality. In D. Liu, C. Dede, R. Huang, & J. Richards (Eds.), Virtual, augmented, and mixed realities in education (pp. 19–33). Smart Computing and Intelligence), Singapore: Springer.

Tasker, R. (2017). Molecular-level visualisation for educational purposes. In M. Springborg & J. Joswig (Eds.), *Chemical modelling* (pp. 261–282). Cambridge: Royal Society of Chemistry.

Treagust, D. F., Chittleborough, G., & Mamiala, T. (2003). The role of submicroscopic and symbolic representations in chemical explanations. *International Journal of Science Education, 25*(11), 1353–1368. https://doi.org/10.1080/0950069032000070306.

Treagust, D. F., Won, M., & Duit, R. (2014). Paradigms in science education research. In N. G. Lederman & S. K. Abell (Eds.), *Handbook of research on science education* (2nd ed., pp. 3–17). New York: Routledge.

Tse, C.-M., Li, H., Leung, K.-S., Lee, K.-H., & Wong, M.-H. (2011). Interactive drug design in virtual reality. Paper presented at the 15th International Conference on Information Visualisation, 2011/07.

Won, M., Yoon, H., & Treagust, D. F. (2014). Students' learning strategies with multiple representations: Explanations of the human breathing mechanism. *Science Education, 98*(5), 840–866. https://doi.org/10.1002/sce.21128.

Wu, H.-K., & Shah, P. (2004). Exploring visuospatial thinking in chemistry learning. *Science Education, 88*(3), 465–492. https://doi.org/10.1002/sce.10126.

Chapter 5
Faculty Motivation for Scholarly Teaching and Innovative Classroom Practice—An Empirical Study

Zakiya S. Wilson-Kennedy, Liuli Huang, Eugene Kennedy, Guoqing Tang, Margaret I. Kanipes and Goldie S. Byrd

Abstract The transformation of undergraduate science and math education will require broad-based faculty motivation to integrate innovative evidence-based pedagogies into classroom practice and advance scholarship in teaching and learning. Currently, the traditional lecture approach remains the dominant model in these disciplines. While research suggests that high-impact practices and active learning pedagogies may improve the success of students from historically underrepresented groups (HUGs), faculty must be motivated to invest their time and effort to do this work. Consequently, three strands of motivation theory (motivation, self-determination, and self-efficacy) may provide a framework for exploring faculty adoption and advancement of potentially transformative practices. Herein, motivation theory can support our investigation of the personal and institutional dynamics that promote and/or hinder faculty integration of high-impact and evidence-based pedagogical innovations into classroom practice. The study reported in this chapter uses quantitative methods to explore faculty engagement in scholarly teaching and other innovative educational practices that purport to increase success of HUGs in the STEM fields at a land-grant doctoral research university in the southeastern region of

Z. S. Wilson-Kennedy (✉)
College of Science, Louisiana State University, Baton Rouge, LA, USA
e-mail: zwilson@lsu.edu

L. Huang · E. Kennedy
School of Education, Louisiana State University, Baton Rouge, LA, USA
e-mail: liulihuang.lsu@gmail.com

E. Kennedy
e-mail: ekennedy@lsu.edu

G. Tang
Department of Mathematics, North Carolina A&T State University, Greensboro, NC, USA
e-mail: tang@ncat.edu

M. I. Kanipes
University Honors Program, North Carolina A&T State University, Greensboro, NC, USA
e-mail: mikanipe@ncat.edu

G. S. Byrd
Wake Forest School of Medicine, Winston Salem, NC, USA
e-mail: gbyrd@wakehealth.edu

M. Schultz et al. (eds.), *Research and Practice in Chemistry Education*,
https://doi.org/10.1007/978-981-13-6998-8_5

the USA. The university has a historical mission of educating HUGs and a designation of HBCU (Historically Black Colleges and Universities). One primary research question guides this effort: "*To what extent do HBCU STEM faculty feel motivated to engage in high-impact, evidence-based pedagogies?*" This work is important in the broader STEM education community for understanding how to foster faculty engagement in transforming the paradigm of instruction in the college classroom with a focus on student success.

Introduction

Within the USA, students exit undergraduate science, technology, engineering, and mathematics (STEM) disciplines at high rates (Wilson et al. 2012a, 2012b; Committee on Prospering in the Global Economy of the twenty-first century (US) and Committee on Science Engineering and Public Policy (US) 2007; Committee on Prospering in the Global Economy of the twenty-first century (US) and Committee on Science Engineering and Public Policy (US) 2010). Students of color from historically underrepresented groups (HUGs) and women leave at higher rates, in most STEM disciplines, than their Asian and Caucasian male peers (National Academy of Sciences, National Academy of Engineering, & Institute of Medicine, 2011).

Numerous reports and national leaders in STEM education have noted the critical role that evidence-based teaching methodologies play in supporting the retention and persistence of *all* students, particularly those HUGs (Handelsman et al., 2004; Wieman, 2012, 2007; National Research Council, 2012; Anderson et al., 2011; President's Council of Advisors on Science and Technology 2012). While some innovative teaching practices (e.g., Math Emporium (Twigg, 2011), SCALE-UP (Gaffney, Richards, Kustusch, Ding, & Beichner, 2008), Process Oriented Guided Inquiry Learning—POGIL (Moog & Spencer, 2008), and Peer Instruction (Mazur & Hilborn, 1997) have gained national attention, the traditional lecture approach remains the dominant model for STEM education at the undergraduate level. While convenient and efficient, the traditional lecture approach may not provide the type of learning environment that promotes the success and persistence of HUGs in STEM academic programs.

The adoption and integration of potentially transformative educational practice into classroom instruction require faculty to be motivated to do this work and to possess the core skills to function in this area of scholarly activity. Notably, faculty who are well trained in their technical areas of research are often less well trained in pedagogy and the implementation of high-impact educational practices.

Further, complex institutional dynamics may impact faculty members' willingness and efficacy in adopting diverse teaching practices. Studies have found that while many faculty members have both an awareness and interest in integrating innovative, evidence-based approaches into their teaching practices (Dancy & Henderson, 2010; Borrego, Froyd, & Hall, 2010; Macdonald, Manduca, Mogk, & Tewksbury, 2005), STEM faculty often work within institutional structures that favor traditional lecture approaches (American Association for the Advancement of Science, 2014). Academic reward systems and competing workload priorities may limit faculty members' ability to focus on teaching; and studies have suggested that time constraints, large class sizes, broad content coverage expectations, and immobile physical facilities are factors that generally work against faculty adoption of broad, systematic pedagogical change (Austin, 2011; Tagg, 2012; American Association for the Advancement of Science, 2014).

Institutional culture can also have a significant impact on faculty's motivation for excellence in teaching (Feldman & Paulsen, 1999). Faculty within research universities often receive many affirmative messages about the priority that they should place on research with greater ambiguity about the priority that they should place on teaching (Wright, 2005; Diamond & Adam, 1998). In these cases, faculty may experience conflict or incongruence between the value that they place on teaching and the value that they perceive that their colleagues and supervising administrators place on teaching (Wright, 2005). Whether or not faculty members are motivated to integrate innovative, evidence-based strategies into their teaching are related to the real or perceived institutional support for teaching. Understanding faculty motivation and the climate for teaching excellence are important contextual factors to consider when investigating faculty engagement in pedagogical innovation and classroom practice.

Theoretical Framework

Three strands of motivation theory (intrinsic and extrinsic motivation, self-determination and social support, and self-efficacy) comprise the framework used to investigate the personal and social dynamics contributing to faculty interest and engagement in STEM pedagogical innovations. This study was closely modeled after the work of Hardré, Beesley, Miller, and Pace (2011) which investigated faculty motivation toward research.

Studies have demonstrated that **intrinsic and extrinsic motivation** is predictors of personal action (Harackiewicz, 2000; Pintrich &Schunk, 2002; Deci & Flaste, 1995; Hardré et al., 2011). Intrinsic motivation occurs when an individual participates in an activity because of an innate interest or enjoyment in the activity. Accordingly, intrinsic motivation is a well-established predictor of an individual's willingness to expend effort, engagement, and achievement on a desired task (Harackiewicz, 2000; Pintrich & Schunk, 2002; Deci & Flaste, 1995; Hardré et al., 2011). Intrinsic motivation is key for innovation. When external pressure contributes to personal action, extrinsic motivation predicts minimal effort and a hesitancy to take risks or innovate (Harackiewicz, 2000; Pintrich & Schunk, 2002; Deci & Flaste, 1995; Hardré et al., 2011).

Self-determination theory (SDT), a construct contributing to motivation, is a meta-theory that helps to explicate how social and cultural factors impact an individual's drive and sense of volition (self-determination theory (SDT)). Herein, autonomy, competence, and relatedness contribute to the conditions wherein an individual's performance, persistence, and creativity are optimized. SDT suggests that faculty autonomy predicts work effort and performance (Deci & Ryan, 1987, 2002a, 2002b). Individuals' perceptions of competency (and expertise) support their motivation to engage in work-related activities (Ryan & Deci, 2000). Moreover, relatedness, as evidenced by the "interpersonal" support of authority figures (campus leadership), has been demonstrated to predict job performance and satisfaction (Deci & Ryan, 2002b).

Self-efficacy espouses an individuals' confidence to engage in tasks or activities despite obstacles and challenges (Bandura, 1997; Reeve, 1996). It has been shown to be a predictor of scholarly productivity and collaboration among faculty (Hardré et al., 2007; Feldman & Paulsen, 1999).

While several studies have used intrinsic motivation, self-determination, and self-efficacy to investigate faculty motivation as predictors of productivity in research (Hardré et al., 2007, 2011; Hardré 2012), few studies have applied motivation theory to investigate faculty engagement in innovative pedagogical reformation. These constructs can be used to investigate faculty proclivity to lead and collaborate on integrating evidence-based pedagogies in the classroom.

Faculty perceptions of their organization's valuation of teaching versus other scholarly work may impact how faculty members expend effort within the academy. Accordingly, a study of faculty motivation, self-determination, and self-efficacy for teaching innovation must also include an investigation of faculty members' perceptions of the institutional priorities and environment for teaching. Herein, these theories may explain why faculty seek out professional development in teaching, seek membership in supportive communities, and engage in innovative pedagogical practices.

Overview of the Research Study

The transformation of undergraduate STEM education will require broad-based faculty motivation to integrate innovative, evidence-based pedagogies into classroom practice. Sustainable change from traditional lecture approaches to more integrative high-impact practices will occur when faculty members are motivated to change this. The study reported in this chapter uses quantitative methods to explore faculty motivation for engaging in scholarly teaching and other innovative educational practices that purport to increase the success of HUGs in the STEM fields. Herein, three strands of motivation theory (intrinsic and extrinsic motivation, self-determination, and self-efficacy) provided a framework for exploring the personal and institutional dynamics that promote and/or hinder faculty integration of evidence-based pedagogical innovations into classroom practice (Colbeck, Cabrera, & Marine, 2002) at a land-grant doctoral research university in the southeastern region of the USA. The university has a historical mission of educating HUGs and a designation of HBCU (Historically Black Colleges and Universities).

This research study seeks to understand the context in which faculty members engage in scholarly activity. Because of the changing mission of the university from focusing on teaching to one focusing on research, this investigation sought to understand faculty scholarly activities related to teaching versus technical research in their disciplines, their commitment to teaching excellence and engagement in evidence-based teaching practices. The primary research question guiding this study is: *To what extent do HBCU STEM faculty feel motivated to engage in high-impact, evidence-based pedagogies?* This work is important in the broader STEM education community for understanding how to foster faculty engagement in transforming the paradigm of instruction in the college classroom with a focus on student success.

Context of the Study

HBCUs provide a vital institutional context for investigating faculty motivation for improving undergraduate education, particularly in STEM disciplines, for HUGs. HBCUs have a tremendous legacy of providing access to higher education to economically and socially disadvantaged students, particularly HUGs (Mack, Rankins, & Winston, 2011; Goins et al., 2009; Babco, 2003). As an example, HBCUs are only 3% of four-year institutions but produce 20% of African-American scientists and engineers (National Science Foundation, 2013).

In addition to their unique role in US higher education, HBCUs often have complex internal dynamics. As an example, the study was completed within a single institution, whose mission has changed from one focused primarily on teaching to one focused on STEM research (Fig. 5.1). The university led one of the first US National Science Foundation (NSF) Engineering Research Center (ERC) at an HBCU. Research activity is strong throughout the university; however, some STEM units have high numbers of faculty focused on teaching and have limited infrastructure for conducting research as evidenced by facilities, number of graduate students and postdocs, and other inputs normative to research units with high research activity.

The university has a high number of faculty at the associate professor level and a high percentage who are retirement eligible. Further, the university has a post-tenure review process in which faculty must submit evidence of scholarly and service activity on a five-year cycle. For STEM departments with low infrastructure for discipline-based research activity and higher teaching loads, faculty must determine how to pursue scholarly activities. It is within this context that faculty engagement in scholarly teaching has been investigated.

Though the institution has made significant investments in teaching supported through NSF sponsored projects and a STEM Center for Active Learning, a reorganization of the academic structure of the university (which changed the tenure home of many faculty) and changes in institutional and accountability structures confound the messaging about the institutional valuation of teaching and scholarly activity in this area. To this end, this ***HBCU with its Carnegie classification of doctoral higher research, provides a rich institutional context for investigating the impact of faculty***

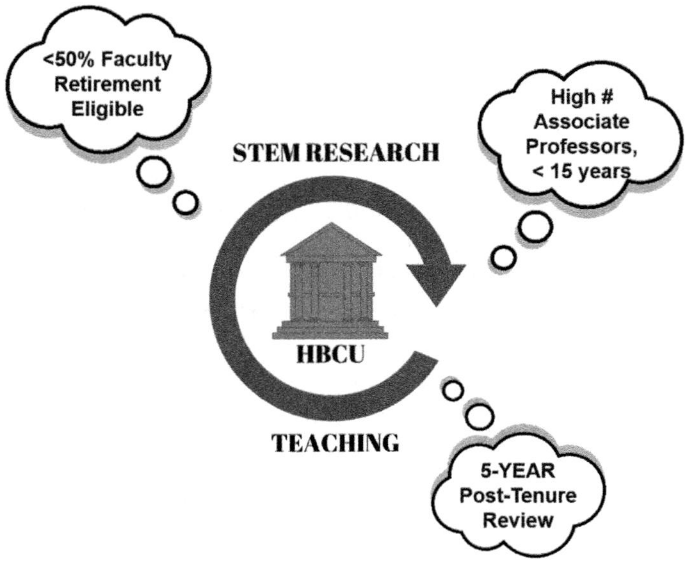

Fig. 5.1 HBCU context for innovative teaching at the study institution

motivation toward scholarly activity in teaching and learning and faculty perceptions of developmental and support structures for the integration of high-impact, evidence-based pedagogies in their teaching.

Methodology

This study employs quantitative methods to investigate faculty members' engagement in pedagogical innovations in undergraduate education at a land-grant doctoral higher research HBCU in the southeastern region of the USA. We define innovative pedagogies as practices beyond traditional lectures which can include Math Emporium (Twigg, 2011), SCALE-UP (Gaffney et al., 2008), Process Oriented Guided Inquiry Learning—POGIL (Moog & Spencer, 2008), Peer Instruction (Mazur & Hilborn, 1997) and other practices such as undergraduate research, team/group work, signature projects, and other high-impact practices.

The primary research question was contextualized into two main inquiry themes, as described in Table 5.1: (1) the organizational context and environment for teaching and (2) the teaching values and perceptions of faculty members. The measures for this study, as provided in Table 5.2, were aligned with sub-questions, which corresponded to either of the themes mentioned above. Given that motivation and self-determination describes an individuals' ideation of an activity based on their personal valuation, enjoyment, and perceptions of support for the said activity, a set of sub-questions were designed to investigate elements of faculty perceptions of how they valued teaching and how they perceive that their colleagues and the institution value teaching. The sub-questions also probe how faculty actually spend their time and looks for alignment between their valuation and their actual efforts. The sub-questions also investigate faculty self-efficacy for engaging in scholarly teaching as they pursue innovative teaching practice.

Descriptive statistics were used to analyze responses collected using the measures comprising the analytical instrument, which responded to the research questions in Table 5.1. This methodology focused on describing the organizational context for teaching and learning in the academic units with a goal of understanding the extent to which a positive valuation for instructional innovation exists at both the organizational and individual faculty levels. Additionally, the study sought to describe strategies for addressing the impact of the project's activities on two key outcomes: faculty engagement in scholarly teaching and use of innovations beyond the traditional lecture in their classes, and faculty attitudes (receptiveness) toward instructional innovations.

Research Measures

The research measures used in this study were modeled after a motivation construct for scholarly productivity in disciplinary research (Hardré et al., 2011). Table 5.2 outlines these measures and includes single item and scales. All of the items were compiled into an online survey which was deployed to faculty.

Participants

Faculty affiliated with current and prior federally funded projects at the HBCU along with its former Academy of Teaching and Learning and the STEM Center for Active Learning were the targets of this study. The project also sought faculty who opted out of participation in these communities, investigating their motivation to pursue and lead pedagogical innovations and their membership in other informal networks focused on scholarly activities in teaching and learning. Participation was voluntary, and the institutional review board approved the study at the HBCU. An online survey was deployed to capture faculty responses. Figure 5.2 presents data on the demographic characteristics of respondents to the survey, which included a total of 68 faculty members. These respondents included tenured and tenure-track faculty at the ranks of assistant, associate, and full professors, as well as instructors and adjuncts, professional staff, and others. The disciplinary areas represented in the study included mathematical and physical sciences, life and agricultural sciences, engineering, technology, social and behavioral science, and arts and humanities. Mathematics and physical science as well as engineering had the highest number

Table 5.1 Alignment of research sub-questions with two primary themes for the study and their corresponding source on the instrument

Research questions	Measures (See Table 5.2)
Organizational context and environment for teaching	
1.1 To what extent do faculty perceive an environment supportive of teaching?	Environment for teaching scale
1.2 To what extent is teaching valued in the organization?	Expected investment
1.3 What is the percentage of effort faculty devote to teaching?	Percentage of effort
	Productivity outcomes
1.4 What is the teaching load of faculty?	Teaching load
Teaching values and perceptions of faculty	
1.5 To what extent do faculty value teaching?	Personal value
1.6 What is faculty "ideal" level of effort focused on teaching?	Ideal investment
1.7 Do faculty have a high sense of efficacy with respect to teaching?	Self-efficacy for teaching
1.8 Does faculty motivation for teaching tend to be extrinsic or extrinsic?	Teaching motivation

Table 5.2 Description of research measures and alignment with research questions and theoretical framework

Research measures: individual items and scales	Research question	Theory
Demographics. Age, gender, ethnicity, family commitment, academic rank, and academic area		
Environment for teaching. Participants indicated their perceptions of departmental and interpersonal support for teaching. A 7-item instrument based on SDT and contextualized from the Work Climate Questionnaire (Baard et al. 2004). Sample items: "My department is socially supportive of teaching excellence, encouraging me to collaborate with other faculty members," "My SOTL activities are valued by my department and my college," and "My department provides me with choices in the pedagogical questions and issues that I investigate." 5-point Likert-type scale, anchors: 1 = "Not at all true"; 3 = "Somewhat true"; 5 = "Very much true" ($\alpha = 0.834$)	1.1	SDT
Expected investment of effort. Indicator of the effort that their department expects them to invest in research, teaching, service, and scholarly teaching. Question: "Based on the functionally defined reward structure in your department, what percentage of your non-financial personal resources (such as time, energy and effort) are you expected to invest in each of the following professional activities?" Percent response constrained to 100%	1.2	IEM, SDT
Percentage of effort. Indicator of the effort invested in research, teaching service, and scholarly teaching. Question: "Based on your own personal standard, how much effort do you invest in each of the following professional activities?"	1.3	
Productivity outcomes. "How many peer-reviewed pedagogical research publications did you count as author or co-author on in the past three calendar years?" and "How many national or international professional pedagogical research conference presentations did you count as author or co-author on in the past three calendar years?" The combined total of publications and presentations will constitute our measure of teaching research productivity	1.3	
Teaching load. Faculty members indicate their annual teaching load in terms of courses per year on average. The question was as follows: "How many courses per academic year do you teach on average?"	1.4	
Personal value. Indicator of the personal value for research, teaching, service, and scholarly teaching. Question: "To what extent do you personally value each these professional activities?"	1.5	IEM
Ideal investment of effort. Participants indicated the percentage of effort that they personally felt should be invested in research, teaching, service, and scholarly teaching. Question: "Based on your own professional standards, how much effort do you invest in the following activities?" Percent response constrained to 100%	1.6	IEM, SDT

(continued)

Table 5.2 (continued)

Research measures: individual items and scales	Research question	Theory
Self-efficacy for teaching. Adapted 24-item self-efficacy for teaching scale (Hardré et al., 2007). Sample items: "Because … teaching is of personal important," "… task of teaching is interesting," "I get paid for it," etc. Responses on a 10-point scale (anchored as: 1 = Low, 10 = High (= 0.98)	1.7	SE
Motivation for teaching. Adapted 15-item instrument (Hardré et al., 2007) examining, "Why do you teach?" Sample items: intrinsic ("I teach because I like to do it"); and extrinsic ("I teach because I have to, to keep my job"). Responses on a 5-point Likert scale, anchored as: 1 = "Not at all true"; 3 = "Somewhat true"; 5 = "Very much true" (intrinsic scale $\alpha = 0.82$; extrinsic scale $\alpha = 0.468$).	1.8	IEM

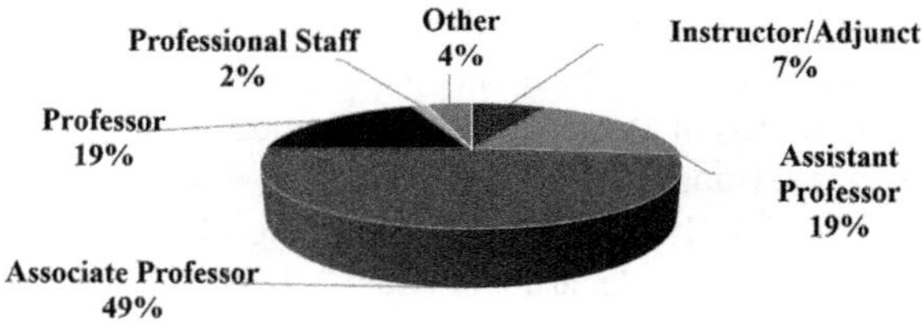

Fig. 5.2 Respondents by faculty status

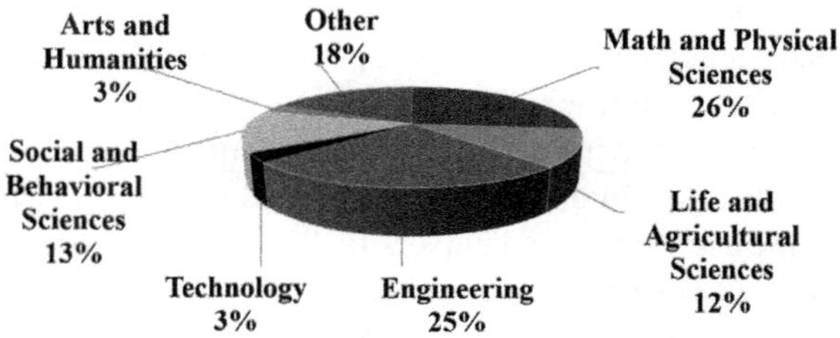

Fig. 5.3 Respondents by academic area

of respondents. The majority of the respondents were associate professors (49%, Fig. 5.2). Tenured faculty (63%) followed by tenure track (24%) and neither on the tenure track nor tenured (13%) responded. 62% of the respondents were male and 38% were female, and they represented several academic areas (Fig. 5.3).

Results and Discussion

University faculty advance their teaching and research within complex institutional dynamics including changing institutional priorities and broadening institutional missions. Herein, faculty motivation to lead and collaborate on scholarly teaching and learning innovations is tied to their sense of empowerment, agency, and their perceptions of how this work is valued within the institution.

Faculty members provided self-assessments of the professional valuations of academic work, namely research in their disciplines, teaching, service, and scholarly teaching as indicated by research on teaching. The instrument for these self-assessments probed the organizational context and environment for teaching, as well as the faculty member's valuation of their teaching values and perceptions.

Organizational Context and Environment for Teaching

Question 1.1 *To what extent do faculty perceive an environment supportive of teaching?* The environment for teaching is an important factor when considering faculty motivation. Notably, the respect of colleagues and general assurances that efforts expended in this area will be rewarded play a role in how faculty spend their time and if they collaborate with colleagues on these efforts. A frequency distribution for items related to respondents' perceptions of their departments' support for teaching is presented in Table 5.3. Most respondents (73.2%) report that they receive encouragement (ratings of 3, 4, or 5) from their department to collaborate with colleagues on teaching (Item 3.1); that scholarly teaching is *somewhat* to *highly* valued (ratings of 3 to 5) in the department (Item 3.2) as well as at the college level (Item 3.3); and that there is a culture of *commitment to excellence* in teaching (Item 3.4). However, more than half report that research is valued more highly than teaching (Item 3.5) and a significant percentage report that their departments provide limited support for excellence in teaching (Item 3.7).

Scale: Environment Supportive of Scholarly Teaching. The question set (Table 5.3) addressed the respondent's perceptions of support for scholarly teaching in the department and university. With the exception of Item 3.5 for this question set, correlations among the remaining seven items were all positive and stated in such a way as to indicate an environment supportive of scholarly teaching. The mean over these items was used to construct a scale. The reliability of the resulting scale was 0.834. Tables 5.4, 5.5, 5.6, and 5.7 summarize the means on this scale, disaggregated by demographic variables, e.g., faculty rank, tenure status, academic division, and gender. These data provide insights into survey respondents' perceptions of the supportiveness of the environment with regard to scholarly teaching. The scale has a range of 1 (limited support for teaching) to 5 (high support for teaching).

With regard to rank, instructors and assistant professors have the highest values, indicating on average more positive perceptions of teaching support than other ranks. Similarly, tenured faculty members report the lowest average when tenure status is considered. The mean for the social and behavioral sciences is higher than other academic departments, and the mean for women respondents is higher than that of men and other.

In Table 5.3, Item Q3.3, respondents were asked if scholarly teaching activities were valued by the university and counted toward tenure. More than half of the participants responded "Somewhat True" to "Very Much True." The majority of respondents indicated that their department was socially support of teaching and that they were encouraged to collaborate with colleagues on teaching. Respondents indicated that the climate of the institution was such that teaching and research on teaching were "Somewhat Valued" to "Highly valued."

Question 1.2 *To what extent is teaching valued in the organization?* Two items on the survey addressed the extent to which teaching was viewed as being valued by the institution. The frequency distributions for these items are presented in Tables 5.8 and 5.9. With regard to the value of teaching as well as the value of research on teaching, the response of most respondents was "somewhat valued."

Question 1.3 *What is the percentage of effort faculty devote to teaching?* Because of its prevalence as the primary indicator of research productivity, we used pedagogical research disseminated as peer-reviewed publications and national/international conference presentations to assess productivity outcomes for scholarly teaching. We asked faculty members to report those publications and presentations for over the past three years. Additionally, the survey asked respondents to indicate the amount

Table 5.3 Environment supportive for scholarly teaching scale ($\alpha = .834$)

	1 "Not at all True"	2	3 "Somewhat True"	4	5 "Very Much True"
Q 3.1 My department is socially supportive of teaching excellence, encouraging me to collaborate with other faculty members on teaching ($n = 56$)	7 (12.5)	8 (14.3)	13 (23.2)	14 (25.0)	14 (25.0)
Q 3.2 My Scholarly teaching activities are valued by my department and my college ($n = 56$)	6 10.7)	8 (14.5)	20 (35.7)	15 (26.8)	7 (12.5)
Q 3.3 Scholarly teaching activities are valued by the university and count toward tenure and promotion ($n = 56$)	7 (12.5)	13 (23.2)	18 (32.1)	11 (19.6)	7 (12.5)
Q 3.4 Within my department, there is a shared commitment to excellence in teaching ($n = 56$)	6 (10.7)	11 (19.6)	17 (30.4)	13 (23.2)	9 (16.1)
Q 3.5 Within my department, research is valued more highly than either teaching or scholarly teaching ($n = 55$)	4 (7.3)	8 (14.5)	15 (27.3)	11 (20.0)	17 (30.9)
Q 3.6 My department provides me with choices in the scholarly teaching questions and issues that I investigate ($n =$ 55)	17 (30.9)	11 (20.0)	10 (18.2)	11 (20.0)	6 (10.9)
Q 3.7 My department provides resources for teaching excellence ($n = 56$)	10 (17.9)	16 (28.6)	20 (35.7)	8 (14.3)	2 (3.6)

Note Values in parentheses are row percentages

Table 5.4 Environment for teaching support, by primary rank

	Mean	N	Standard deviation	Minimum	Maximum
Instructor/Adjunct	3.8167	4	.58468	3.33	4.67
Assistant Professor	3.0556	9	1.20185	1.50	4.67
Associate Professor	2.9222	30	.90944	1.00	4.83
Professor	2.7167	10	.73304	1.67	3.83
Other	2.8889	3	.78764	2.00	3.50
Total	2.9690	56	.91757	1.00	4.83

Table 5.5 Environment for teaching support, by tenure status

	Mean	N	Standard deviation	Minimum	Maximum
Tenure track	3.3472	12	1.07181	1.50	4.83
Tenured	2.7973	37	.82415	1.00	4.50
Neither on tenure track nor tenured	3.2286	7	.99488	1.83	4.67
Total	2.9690	56	.91757	1.00	4.83

Table 5.6 Environment for teaching support, by academic division

	Mean	N	Standard deviation	Minimum	Maximum
Math and physical sciences	3.0289	15	.80118	1.00	4.17
Life and agricultural sciences	3.6389	6	.69456	2.50	4.67
Engineering	2.3810	14	.84335	1.00	3.67
Technology	2.4167	2	.58926	2.00	2.83
Social and behavioral sciences	3.8889	6	1.00370	2.17	4.83
Arts and humanities	2.6667	2	1.17851	1.83	3.50
Other	2.9242	11	.73547	1.67	4.33
Total	2.9690	56	.91757	1.00	4.83

Table 5.7 Environment for teaching support, by gender identity

	Mean	N	Standard deviation	Minimum	Maximum
Male	2.7505	33	.87600	1.00	4.67
Female	3.3636	22	.83355	1.83	4.83
Other	1.5000	1	.	1.50	1.50
Total	2.9690	56	.91757	1.00	4.83

Table 5.8 To what extent is teaching valued in your institution? (5-point Likert scale: 1 = "Not valued"; 3 = "Somewhat valued"; 5 = "Highly Valued")

	Frequency	Percent	Valid percent	Cumulative percent
1 = "Not valued"	1	1.4	1.8	1.8
2	11	15.1	20.0	21.8
3 = "Somewhat valued"	23	31.5	41.8	63.6
4	13	17.8	23.6	87.3
5 = "Highly valued"	7	9.6	12.7	100.0
Total	55	75.3	100.0	

Table 5.9 To what extent is the scholarship of teaching and learning valued in your institution? (5-point Likert scale: 1 = "Not valued"; 3 = "Somewhat valued"; 5 = "Highly Valued")

	Frequency	Percent	Valid percent	Cumulative percent
1 = "Not valued"	3	4.1	5.5	5.5
2	10	13.7	18.2	23.6
3 = "Somewhat valued"	20	27.4	36.4	60.0
4	9	12.3	16.4	76.4
5 = "Highly valued"	13	17.8	23.6	100.0
Total	55	75.3	100.0	

of *effort* they devoted to research, teaching, service, and scholarly teaching. Respondents were instructed to assign percentages such that the total summed to 100%. Descriptive statistics for this item are presented in Fig. 5.4. These results indicate that the majority of effort expended by these respondents is directed toward teaching, followed by research, service, and scholarly teaching. These data also accentuate the importance these respondents placed on teaching, relative to research, service, and research on teaching.

Question 1.4 *What is the teaching load of faculty?* The most common teaching load per semester was 7–9 h and the most common number of courses in which scholarly teaching was integrated was 2–3.

Teaching Values and Perceptions of Faculty

Question 1.5 *To what extent do faculty value teaching?* Respondents ranked their valuation of research, teaching, service, and research on teaching with regard to their *importance to you*. Respondents assigned ranks using a scale of 0–100, with 100 indicating greater importance. Descriptive statistics for this item are presented in Fig. 5.5. These results indicate that teaching and research rank among the most important activities, with research on teaching and service being less important.

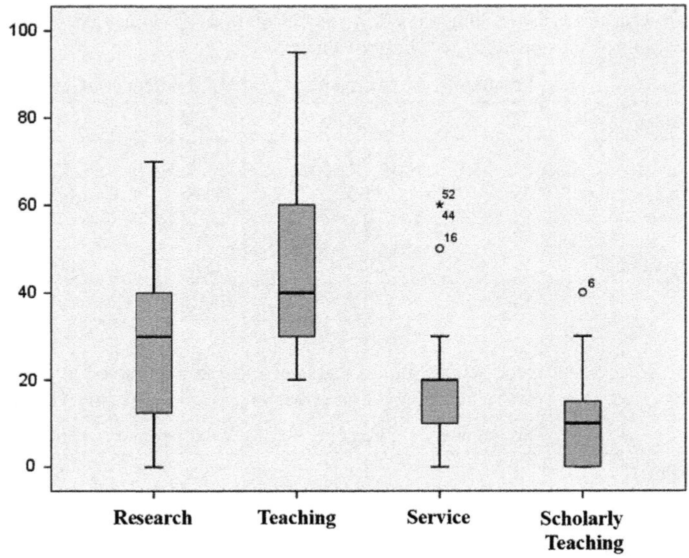

Fig. 5.4 Distribution of effort teaching, research, etc.

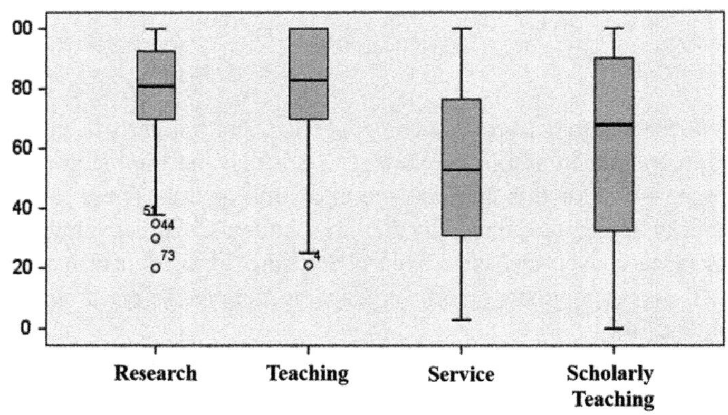

Fig. 5.5 How do you rank the following professional activities with regard to their importance to you?

Question 1.6 *What is faculty "ideal" level of effort focused on teaching?* The survey asked respondents to indicate the effort they are *expected* to devote to research, teaching, service, and research on teaching and to indicate their *ideal* investment of effort. For both items, respondents were instructed to assign percentages such that the totals summed to 100%. The results for expected effort are presented in Fig. 5.6.

The results for ideal effort are presented in Fig. 5.7. The research for research and teaching are similar with respect to expected effort, while the median ideal effort for

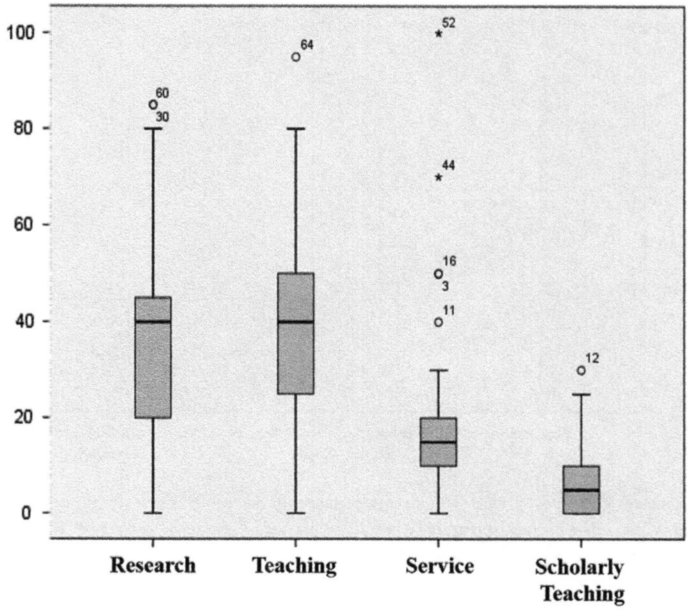

Fig. 5.6 Based on the functionally defined reward structure in your department, what percentage of your non-financial personal resources (e.g., time, energy, effort) are you expected to invest in each of the following professional activities?

teaching is higher than that of research. For both items, several extreme responses are notable in the box plots, indicating individuals whose percentages are outside of the norm.

Question 1.7 *Do faculty have a high sense of efficacy with respect to teaching?* To address this question, items related to the self-efficacy scale on the survey were analyzed (Table 5.10). The correlations among these items were positive, and the estimated coefficient alpha reliability was .98. The mean over these items was used to construct a scale. The range for this scale was 1 (low self-efficacy) to 10 (high self-efficacy). On a scale of 1–10, the mean for the respondents on the self-efficacy scale is 5.9. This result indicates an average slightly higher than the midpoint of the range of the scale. This moderate value is likely a target for improvement.

Question 1.8 *Does faculty motivation for teaching tend to be extrinsic or extrinsic?* To address this question, items on the survey (Table 5.11) that address motivation for teaching were divided into two groups based on their content. Intrinsically worded items reflected positive views about teaching and the ability of the respondent to be effective in the classroom. Items 1 through 8 on this portion of this survey were placed in this category. These items had positive correlations with each other and the estimated coefficient alpha reliability was .82. Items that reflect extrinsic attitude toward teaching were items 9 through 12 on this portion of the survey. These items also had positive correlations with each other and negative or zero correlation with

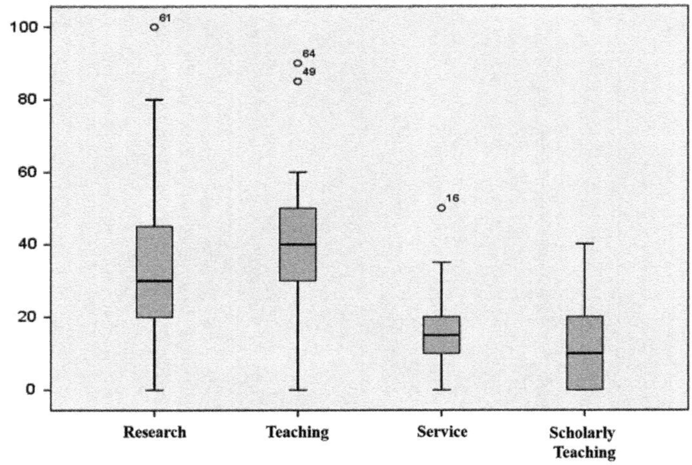

Fig. 5.7 Based on your own professional standards, what do you believe would be the ideal reward structure for your current institution? That is, what percentage of your non-financial personal resources (e.g., time, energy, effort) should you invest in each of the following professional activities?

items in the intrinsic teacher motivation scale. However, the estimated reliability is low at .468. This scale reflects the mean over the 4 items. They indicate that the average on both the intrinsic and extrinsic scales were slightly above the midpoints of the scale, indicating that respondents did not hold very strong views in either dimension.

Conclusions and Implications for Practice

The primary research questing guiding this study is, "*To what extent do HBCU STEM faculty feel motivated to engage in high-impact, evidence-based pedagogies?*" Herein, three strands of motivation theory (intrinsic and extrinsic motivation, self-determination, and self-efficacy) comprise the framework for exploring how institutional culture and personal values impact faculty resource allocations for teaching and scholarly activity related to research on teaching in an HBCU with a changing mission from primarily teaching to research.

Our findings suggest that faculty at all levels within the STEM disciplines at the doctoral research HBCU, under study, indicated that the university has an environment that is supportive of teaching. Faculty in the non-tenured ranks were more likely to perceive that the institutional culture was supportive of teaching than tenured faculty. Faculty within the Life and Agricultural Sciences and within the Social and Behavioral Science were more likely to perceive support for teaching than any of the other STEM divisions (e.g., math and physical sciences; engineering; or technol-

Table 5.10 How Confident you are in performing scholarship of teaching and learning (or SoTL) tasks ($\alpha = 0.98$)

	N	Min	Max	Mean	Standard deviation
Writing SOTL journal articles for peer review	57	1	10	5.56	2.916
Delivering SOTL conference papers and presentations	57	1	10	5.82	3.163
Supervising students' SOTL research projects	57	1	10	5.74	3.068
Analyzing SOTL data	57	1	10	5.89	2.920
Collecting SOTL data	57	1	10	6.07	2.939
Designing SOTL research projects	57	1	10	5.74	2.967
Applying for SOTL research grants	57	1	10	5.28	2.858
Preparing a SOTL research budget	57	1	10	4.95	2.862
Adhering to SOTL research ethics requirements	57	1	10	6.46	3.071
Collaborating with colleagues about SOTL research	57	1	10	6.47	2.928
Generating SOTL research ideas	57	1	10	6.46	3.224
Reviewing literature for a SOTL research project	56	1	10	6.48	3.145
Keeping up with SOTL research literature	56	1	10	6.39	3.085

ogy). Notably, the majority of faculty respondents indicated that their departments were socially supportive of teaching excellence, with 73.2% agreeing that this was very much true, true, or somewhat true. A significant number also indicated that their departments were supportive of their scholarly teaching activities, with 75% agreeing that this was very much true, true, or somewhat true. To a lesser extent, faculty respondents shared that scholarly teaching had value in tenure and promotion decisions, with 64.2% agreeing that this was very much true, true, or somewhat true.

Additionally, faculty respondents indicated that limited supports were provided for teaching excellence, with 46.5% disagreeing about receiving this type of support. Together, these responses suggest a complicated mix of messages received by faculty members. In most departments, the teaching culture is collaborative, and herein, faculty indicated that there is significant social value among colleagues for engaging in quality teaching activities. However, the university's valuation of teaching is unclear to faculty, as evidenced by the variance in perceived roles in tenure and promotion decisions among different departments.

Table 5.11 Motivation and self-efficacy: by rank

		Intrinsic motivation	Extrinsic motivation	Self-efficacy SoTL
Instructor /adjunct	Mean	4.5000	2.5000	3.5192
	N	4	4	4
	Standard deviation	.75691	.84163	3.78379
Assistant professor	Mean	4.5139	1.9722	6.8889
	N	9	9	9
	Standard deviation	.44828	.68971	1.57932
Associate professor	Mean	4.3458	2.3167	5.9469
	N	30	30	30
	Standard deviation	.54898	.48660	2.71730
Professor	Mean	4.5455	2.2727	6.1538
	N	11	11	11
	Standard deviation	.55417	.51786	3.25558
Other	Mean	4.4167	1.4167	5.3590
	N	3	3	3
	Standard deviation	.47324	.72169	2.21123
Total	Mean	4.4254	2.2193	5.9342
	N	57	57	57
	Standard deviation	.53497	.59212	2.75765

Our findings also suggest that faculty have a strong motivation to teach but may experience tension between what they perceive are institutional priorities and their own personal realities in how they allocate their time and other tangible resources on academic work. Specifically, faculty may be spending more time on teaching when they perceive that they should be spending more time on research. The survey instrument employed for this study investigated what faculty believed should be the ideal way that they invest their time, how they perceive the university expects them to spend their time, and how they actually do spend their time. The investment of time and other tangible resources can be summarized in the following terms: (a) what do you think you should be doing; (b) what do others think that you should be doing; and (c) what are you actually doing. For these questions, faculty ranked the importance of four categories of academic work: (1) research in their discipline; (2) teaching; (3) service; and (4) scholarly teaching or research on teaching and learning.

For each of these, faculty members indicated the priorities for each academic work category across a continuum of 1 for low importance to 4 for high importance. Accordingly, 1.0 was deemed to have a very low importance; 1.5–2.0 corresponded to low importance; 2.5 was the neutral category; 3.0–3.5 corresponded to a high category; and 4.0 corresponded to a very high importance. By a wide margin, faculty indicated that they spent a large amount of effort on teaching, followed by research, service, and then scholarly teaching. When ranking the importance of these academic activities, faculty were more likely to indicate that teaching was more important to them than research; however, research had a high importance that was nearly comparable to most respondents as teaching. However, to a large extent, faculty indicated an institutional expectation that they would spend the majority of their time on discipline-based research and little to no time on research on teaching and learning. Nevertheless, faculty respondents overwhelmingly indicated that they would ideally spend most of their time on teaching, though many indicated strong idealist interests in investing their time into research.

Our findings illustrate the wide variance of responses for faculty perceptions of academic work. When assessing the median values for each category, allocations for both the ideal personal investment and institutions preferred investments corresponded primarily to research and teaching, followed by service and scholarly teaching, in that order. Despite the higher institutional expectation that faculty would spend the majority of their time on research, the respondents indicated that they actually spend the majority of their time on teaching and ideally would spend most of their time on teaching. This suggests a real tension in academic work that faculty are navigating daily on a real-time basis.

Notably, our findings did not indicate that respondents placed a high value on scholarly teaching. While the institutional culture supports teaching excellence, it is not clear that scholarly teaching is highly valued either by the institution or the faculty. This may be due to moderate faculty self-efficacy in leading, designing, and implementing these types of projects.

Several questions on the survey instrument probed faculty's self-efficacy in conducting scholarly teaching. These questions ranged from the project design, to publishing papers, collaborating, evaluating others work, and more. Respondents indicated only moderate self-efficacy on measures associated with scholarly teaching. Stronger support would be needed to cultivate faculty self-efficacy in these areas if the institutional leaders and other faculty desire to see more growth in faculty productivity, confidence, and valuation in scholarly teaching. Given the significant number of faculty who are at the associate professor and full professor ranks, who have spent several years primarily focused on teaching, the empowerment of these faculty members as researchers on teaching and learning can potentially produce institutional benefits on student success and retention, while revitalizing faculty engagement in research activity.

Faculty respondents had the option to provide anonymous perspectives on teaching innovation in the university. Many shared their own personal value on innovative teaching and what they perceive to be the institutional value on scholarly teaching. Respondents indicated a wide range of perspectives. One suggested that teaching and research are both undervalued by the university leadership. Another suggested that the university priority on scientific research is negatively impacting teaching excellence and student outcomes. Several indicated a need for greater institutional support for both discipline-based research and teaching innovation and scholarship.

In summary, responses on the survey demonstrated moderate to strong institutional support for teaching, a strong motivation to teach among the faculty, and a high level of effort that faculty are spending on teaching. However, the university has a much higher valuation for research with fewer tangible supports for teaching in general and very limited support for scholarly teaching, or research on teaching and learning. With only moderate self-efficacy among faculty for leading and collaborating on research on teaching and learning, this area of scholarship represents a potential growth area for the university.

The university under study has experienced an evolving institutional focus, which has transitioned from primarily teaching to research. In transitions such as these, an important realm of scholarship that may be missed is scholarly teaching. When faculty who have a personal motivation to teach are confronted with institutional pressures to increase scholarly activity, they may be motivated to pursue scholarly teaching. Herein, this type of scholarship could advance student success while providing quality research to improve the university for all stakeholders. However, faculty may require more support for engaging in this type of scholarly work.

Implications for Practice

University faculty often receive conflicting message on the important of innovative and scholarly teaching within their institution. National calls for transforming STEM education and improving student success outcomes must take into consideration the environments and conditions in which faculty members work. Beyond providing professional development for faculty to learn about new pedagogies and strategies for advancing scholarly efforts through innovative practice, stakeholders need to develop supportive environments for faculty to advance teaching and learning while facilitating greater student success.

Acknowledgements This work was supported by the US National Science Foundation (NSF) by Award number HRD-1719498.

References

American Association for the Advancement of Science. (2014). Achieving systemic change: A sourcebook for advancing and funding undergraduate STEM Education. In C. Fry (Ed.),. Association of American Colleges and Universities.

Anderson, W. A., Banerjee, U., Drennan, C., Elgin, S., Epstein, I., Handelsman, J., et al. (2011). Science education. Changing the culture of science education at research universities. *Science (New York, NY)*, 331(6014).

Austin, A. E. (2011). Promoting evidence-based change in undergraduate science education. Ann E. Austin.

Baard, P. P., Deci, E. L., & Ryan, R. M. (2004). Intrinsic need satisfaction: a motivational basis of performance and weil-being in two work settings 1. *Journal of Applied Social Psychology, 34*(10), 2045–2068.

Babco, E. (2003). Trends in African American and Native American participation in STEM higher education. *Commission on Professionals in Science and Technology*, 1200.

Bandura, A. (1997). Self-efficacy: The exercise of control. Macmillan.

Borrego, M., Froyd, J. E., & Hall, T. S. (2010). Diffusion of engineering education innovations: A survey of awareness and adoption rates in US engineering departments. *Journal of Engineering Education, 99*(3), 185–207.

Colbeck, C. L., Cabrera, A. F., & Marine, R. J. (2002). Faculty motivation to use alternative teaching methods.

Committee on Prospering in the Global Economy of the twenty-first century(U.S.), & Committee on Science Engineering and Public Policy (U.S.). (2007). Rising above the gathering storm: Energizing and employing America for a brighter economic future. Washington, D.C.: National Academies Press.

Committee on Prospering in the Global Economy of the twenty-first century(U.S.), & Committee on Science Engineering and Public Policy (U.S.). (2010). Rising above the gathering storm revisited: Rapidly approaching category 5. Washington, DC: National Academies Press.

Dancy, M., & Henderson, C. (2010). Pedagogical practices and instructional change of physics faculty. *American Journal of Physics, 78*(10), 1056–1063.

Deci, E. L., & Flaste, R. (1995). Why we do what we do. Putnam Publishing Group.

Deci, E. L., & Ryan, R. M. (1987). The support of autonomy and the control of behavior. *Journal of Personality and Social Psychology, 53*(6), 1024.

Deci, E. L., & Ryan, R. M. (2002a). Handbook of self-determination research. University Rochester Press.

Deci, E. L., & Ryan, R. M. (2002b). The paradox of achievement: The harder you push, the worse it gets.

Diamond, R. M., & Adam, B. E. (1998). Changing Priorities at Research Universities, 1991–1996. Based on: The National Study of Research Universities on the Balance between Research and Undergraduate Teaching (1992), by Peter J. Gray, Robert C. Froh, Robert M. Diammond. ERIC.

Feldman, K. A., & Paulsen, M. B. (1999). Faculty motivation: The role of a supportive teaching culture. *New directions for teaching and learning, 1999*(78), 69–78.

Gaffney, J. D., Richards, E., Kustusch, M. B., Ding, L., & Beichner, R. J. (2008). Scaling up education reform. *Journal of College Science Teaching, 37*(5), 48.

Goins, G., White, C., Foushee, D., Smith, M., Whittaker, J., & Byrd, G. (2009). HBCUs Model for Success. *Successful Models for Effectively Retaining and Graduating Students at North Carolina A&T State University, New York: Thurgood Marshall College Fund Book*.

Handelsman, J., Ebert-May, D., Beichner, R., Bruns, P., Chang, A., DeHaan, R., et al. (2004). Scientific teaching. *Science, 304*(5670), 521–522.

Harackiewicz, J. M. (2000). Intrinsic and extrinsic motivation: The search for optimal motivation and performance. Academic Press.

Hardré, P., Miller, R., Beasley, A., Pace, T., Maxwell, M., & Xie, K. (2007). What motivates university faculty members to do research?: Tenure-track faculty in research-extensive universities. *Journal of the Professoriate, 2*(1), 75–99.

Hardré, P. L. (2012). Community college faculty motivation for basic research, teaching research, and professional development. *Community College Journal of Research and Practice, 36*(8), 539–561.

Hardré, P. L., Beesley, A. D., Miller, R. L., & Pace, T. M. (2011). Faculty motivation to do research: Across disciplines in research-extensive universities. *Journal of the Professoriate, 5*(1), 35–69.
Macdonald, R. H., Manduca, C. A., Mogk, D. W., & Tewksbury, B. J. (2005). Teaching methods in undergraduate geoscience courses: Results of the 2004 On the Cutting Edge survey of US faculty. *Journal of Geoscience Education, 53*(3), 237–252.
Mack, K. M., Rankins, C. M., & Winston, C. E. (2011). Black women faculty at historically black colleges and universities: Perspectives for a national imperative. *Diversity in Higher Education, 11,* 149–164.
Mazur, E., & Hilborn, R. C. (1997). Peer instruction: A user's manual. *Physics Today, 50,* 68.
Moog, R. S., & Spencer, J. N. (2008). *Process oriented guided inquiry learning (POGIL).* USA: Oxford University Press.
National Academy of Sciences, National Academy of Engineering, & Institute of Medicine. (2011). Expanding Underrepresented Minority Participation: America's Science and Technology Talent at the Crossroads. Washington DC: National Academies Press. Available at: http://www.nap.edu/catalog/12984.html.
National Research Council. (2012). Discipline-based education research: Understanding and improving learning in undergraduate science and engineering. National Academies Press.
National Science Foundation, N. C. f. S. a. E. S. (2013). Women, Minorities, and Persons with Disabilities in Science and Engineering: 2013. Special Report http://www.nsf.gov/statistics/wmpd/. Accessed 304 NSF 13.
Pintrich, P. R., & Schunk, D. H. (2002). Motivation in education: Theory, research, and applications. Prentice Hall.
President's Council of Advisors on Science and Technology. (2012). Engage to Excel: Producing One Million Additional College Graduates with Degrees in Science, Technology, Engineering, and Mathematics Executive Office of the President. Available at: http://www.whitehouse.gov/sites/default/files/microsites/ostp/pcast-executive-report-final_2–13-12.pdf.
Reeve, J. (1996). Motivating others: Nurturing inner motivational resources. Allyn & Bacon.
Ryan, R. M., & Deci, E. L. (2000). Self-determination theory and the facilitation of intrinsic motivation, social development, and well-being. *American Psychologist, 55*(1), 68.
Self Determination Theory (SDT). http://selfdeterminationtheory.org/theory. Accessed 21 October, 2018.
Tagg, J. (2012). Why does the faculty resist change? *Change: The Magazine of Higher Learning, 44*(1), 6–15.
Twigg, C. A. (2011). The math emporium: Higher education's silver bullet. *Change: The Magazine of Higher Learning, 43*(3), 25–34.
Wieman, C. (2007). Why not try a scientific approach to science education? *Change: The Magazine of Higher Learning, 39*(5), 9–15.
Wieman, C. (2012). Applying new research to improve science education. *Issues in Science and Technology*, 1–8.
Wilson, Z. S., Holmes, L., Sylvain, M. R., Batiste, L., Johnson, M., McGuire, S. Y., et al. (2012a). Hierarchical mentoring: A transformative strategy for improving diversity and retention in undergraduate stem disciplines. *Journal of Science Education and Technology, 21*(1), 148–156.
Wilson, Z. S., Iyengar, S. S., Pang, S.-S., Warner, I. M., & Luces, C. A. (2012b). Increasing access for economically disadvantaged students: The NSF/CSEM & S-STEM programs at Louisiana State University. *Journal of Science Education and Technology, 21*(5), 581–587.
Wright, M. (2005). Always at odds? Congruence in faculty beliefs about teaching at a research university. *The Journal of Higher Education, 76*(3), 331–353.

Chapter 6
Lessons Learnt from Teaching and Learning During Disruptions

Marietjie Potgieter, Lynne A. Pilcher, Rethabile R. Tekane, Ina Louw and Lizelle Fletcher

Abstract Student protests that are associated with disruption of teaching activities occur sporadically in many countries. Such events usually have a negative impact on teaching and learning, but also provide the impetus for experimentation and renewal of practices. During the student led protest movement #FeesMustFall in South Africa in 2016, three campuses of the University of Pretoria were closed for normal activities for an extended period and most teaching activities went online almost overnight. This study reports on insights gained about the affordances and weaknesses of the online learning environment using self-report data from both lecturers and students. A thematic approach was used to analyse the lecturers' reflections on challenges faced and lessons learnt. The findings exemplify the resilience of lecturers as they embraced blended learning and their willingness to experiment and improve their practice. A mixed-methods approach was used to explore the student experience of the blended learning offering of first-year organic chemistry. The sample was stratified based on prior performance to uncover possible differences between subgroups. The pre- and post-chemistry e-learning surveys, comprising open and structured response questions, were completed by 1166 and 565 respondents, respectively. Inductive analysis of free response data generated themes on the student experience. A negative correlation was found between preference for online learning and performance, which is a reason for concern. The results offer useful pointers for instructional design of

M. Potgieter (✉)
Department of Chemistry, University of Pretoria, Pretoria, South Africa
e-mail: marietjie.potgieter@up.ac.za

I. Louw
Department for Education Innovation, University of Pretoria, Pretoria, South Africa
e-mail: ina.louw@up.ac.za

L. A. Pilcher · R. R. Tekane
Department of Chemistry, University of Pretoria, Pretoria, South Africa
e-mail: lynne.pilcher@up.ac.za

R. R. Tekane
e-mail: reeh.tekane@up.ac.za

L. Fletcher
Department of Statistics, University of Pretoria, Pretoria, South Africa
e-mail: lizelle.fletcher@up.ac.za

M. Schultz et al. (eds.), *Research and Practice in Chemistry Education*,
https://doi.org/10.1007/978-981-13-6998-8_6

blended organic chemistry courses, but they also raise questions about equity and students' metacognitive awareness in a virtual learning environment.

Introduction

Historically, many universities have had to cope with disturbances, disruptions and distractions caused by student protest movements (Ibrahim, 2010; Brooks, 2016). Students have been agents of change in many contexts, frequently for the better. Student protests at the University of Pretoria (UP) began early in 2015 because of rising tuition fees, which placed tertiary education beyond the reach of poor and middle class students. This situation was exacerbated by the inability of the state loan and bursary agency to meet its commitments to students. Protests erupted again in September 2016 after the national announcement that fee increases would be allowed for 2017.

The message was clear that the costs of higher education were too high and unaffordable for the majority of poor black students. The #FeesMustFall movement quickly spread to all public higher education institutions. However, public support for the movement waned when protests became violent and resulted in damage to property. Many institutions closed their campuses for extended periods to prevent injury and further damage to property (Isilow, 2016).

Much has been written on the impact of protests on universities or countries and the drivers for these protests (Tekane, Louw, & Potgieter, 2018), which range from national issues, such as protest against a prominent political figure or political regime (Lewis, 2016), to very specific contextual issues, such as tuition fee increases (Marin, 2012) as in the recent case in South Africa. This chapter will report on the impact of the protests on students' experiences of learning during the disruptions, lecturers' experiences of teaching during student disruptions and the insights derived from these experiences.

Institutional Context

The University of Pretoria (UP) did not escape from the successive waves of student protests in South Africa. The main campus was closed for brief periods in October 2015, and January and February 2016. From 19 September 2016, classes and normal teaching activities ceased. The situation remained uncertain as University authorities attempted to negotiate with leaders of the protest action. As it became clear that no agreement could be reached, the university closed three of its five campuses on 11 October 2016 for the rest of the academic year.[1] To honour its contract with students, the university committed to complete the academic year despite campus closure.

[1] In South Africa, the academic year stretches from February to November of a calendar year.

This meant that the university switched overnight from predominantly face-to-face (F2F) to online (OL) teaching and learning, just after the start of the fourth quarter of the academic year. To accommodate the special needs of the disciplines in two faculties, engineering and science, small numbers of students were granted campus access every day, but student numbers were carefully controlled and contact time was reduced substantially. This arrangement meant that lecturers had to change their planned instruction at short notice in a situation that was too fluid to make provision for proper planning and preparation. Laboratory sessions and tutor classes were either cancelled or replaced with assignments. All midterm sit-down assessments were cancelled, and final examinations were restricted to two hours rather than the standard three-hour duration.

The decision to switch to online teaching and learning was taken in the context that the Council of the University of Pretoria had approved a hybrid teaching and learning model in 2014. As a result, the institution had invested resources in capacity development (hardware, software and staff training) to promote the use of the OL environment beyond technology-enhanced, F2F teaching. Furthermore, UP had a mature learning management system and had embarked on a progressive roll out of free Wi-Fi on campus and in residences. However, at the time of campus closure, lecturers could not assume that all students would be able to access OL resources as students no longer had access to free Internet on campus, computer laboratories or the university library. Data costs in South Africa are high and free Wi-Fi is limited. In recognition of the fact that lack of resources would disadvantage the most vulnerable students, the university set up computer laboratories bordering on the campus and negotiated with mobile network providers to provide free access to specific Internet domains. Lecturers responded to the challenge by implementing a variety of alternative teaching technologies, modes of communication and assessment strategies they had not used before.

Context for This Study

The abrupt switch from face-to-face to (almost) fully online posed a unique opportunity to capture lecturers' and students' experiences of two very different modalities for science teaching and learning. The exceptional circumstances of student protests and the university's response to that presented itself as a quasi-experimental design, something that is rarely possible in education research. In this paper, we present an analysis of lecturers' reflections on lessons learnt about teaching under exceptional circumstances as the context for a report on students' preferences for F2F or OL learning, their experience of both modalities and their performance in an introductory organic chemistry course.

Literature Review

The advantages and disadvantages of fully F2F learning and fully OL learning are well described (Renner, Laumer, & Weitzel, 2014). Traditional F2F learning gives lecturers the opportunity to (i) control the learning environments and (ii) use various teaching methods to cater for students' needs (Acton, Scott, & Hill, 2005). Traditional F2F learning offers a "rich social context" for learners where they are able to interact with one another and also obtain immediate feedback regarding concepts they do not understand (Acton et al., 2005; Hill, Chidambaram, & Summers, 2013). Traditional F2F learning is, however, costly (Renner et al., 2014), not flexible (Hill et al., 2013) and intimidating specifically for those students who are not confident to engage in classroom discussions or ask questions. OL learning, on the other hand, offers flexible, convenient learning environments in which students take ownership of their learning (Renner et al., 2014). However, in OL learning, the lack of F2F, student-to-student or lecturer-to-student interaction, often results in feelings of isolation (Hill et al., 2013).

Blended learning combines the traditional F2F learning with OL learning, hence creating a blend of learning experiences. Furthermore, the optimal combination of these two learning modalities can mitigate their respective weaknesses. The implementation of blended learning in higher education has rapidly grown over the last decade (Graham, Woodfield, & Harrison, 2013, Alammary, Sheard, & Carbone, 2014; Dziuban, Graham, Moskal, Norberg, & Sicilia, 2018), and according to the 2017 New Media Consortium Horizon Report, blended learning is one of the major forces driving the adoption of technology in higher education (Adams Becker et al., 2017). Although the use of blended learning has rapidly grown, there is still no universally accepted definition ascribed to this term (e.g. Brown, 2016).

Blended learning can be categorised into four types, namely *replacement, supplemental, emporium* and *buffet* (Twigg, 2003). Under normal circumstances, most of the lecturers at UP adopted the supplemental type of blending that entailed retaining normal class time meetings and making additional learning resources available online. However, during the 2016 student unrest, lecturers had to implement the replacement type of blend because F2F lectures were substantially reduced and restricted to selected courses. The replaced lectures are often presented online in the form of videos of either a lecturer delivering the lecture or PowerPoint slides with voiceover narration (Twigg, 2003). The sudden implementation of the replacement model meant that both the lecturers and students were forced to change approaches to teaching and learning. Students had to take ownership for their learning and had to take charge of the time, place, setting, path and pace for their learning to occur (Ossiannilson, 2017). In the emporium model, there are no class meetings and students work solely online but within a "learning resource centre", and in the buffet model students can chose from a menu of F2F and online learning activities a combination that suits their own objectives and style.

According to studies on students' perceptions of blended learning, students valued the OL components of blended learning because they improved their analytical skills

(Chen & Jones, 2007) and were convenient, easy to access and promoted autonomy of learning (Akkoyunlu & Soylu, 2008; Chandra & Fisher, 2009). Chemistry undergraduate students indicated that they enjoyed the OL components of the flipped type of blended learning because they promoted self-pacing (Reid, 2016; Christiansen, 2014); provided flexibility; that is, students could study at their own time and place (Smith, 2013; Yeung & O'Malley, 2014); structured their work; and helped them better prepare for examinations and assignments (Flyn, 2015). Students also valued the F2F component as an important part of blended learning because instruction is clearer (Chen & Jones, 2007), they are given the opportunity to ask questions, and they receive immediate feedback (Akkoyunlu & Soylu, 2008; Chandra & Fisher, 2009). The fact that students consider both F2F and OL modalities essential for their learning suggests a need for research-informed implementation of blended learning to optimise the contribution of both modalities.

Part 1: Lecturer's Reflections on Teaching During Campus Closure

Research Design, Data Collection and Analysis

Sir Winston Churchill is often quoted (and disputed) to have said "*Never let a good crisis go to waste*". In this spirit of capitalising on adversity, an opportunity was created for lecturers of the Faculty of Natural and Agricultural Sciences (NAS) at UP to share their reflections on what they have learnt from their experience during the disruptions of 2016. Fifteen lecturers from 13 different departments in agricultural, biological, physical and mathematical sciences participated in a *pecha kucha* event (Beyer, 2011) in February 2017. *Pecha Kucha* is a short presentation style ideally suited to multiple-speaker seminar events. In our case, each presenter gave a 5-minute PowerPoint presentation and the time allocation was carefully managed. Pseudonyms, P1–P15, are used to protect the anonymity of the participants and their departments. The study aimed to address the following research question: What are the lecturer's perceptions of teaching during the 2016 student unrest?

All presentations were video recorded, transcribed and coded thematically for themes on lecturers' lived experiences and their reflections on the effectiveness of their innovations. Individual open coding by two of the researchers followed by comparison and discussion generated a set of themes and sub-themes, which were checked subsequently by a third independent researcher for refinement. The credibility of the findings was also established via peer debriefing (Lincoln & Guba, 1989; Creswell, 2009). The study obtained ethical clearance from the ethics committee of the University of Pretoria (EC180531-191).

Findings

The six themes with selected quotes are presented below.

Theme 1: Restricted Access to Campus Impacts Negatively on Student Learning

The consensus amongst all presenters was that the drastic reduction in contact with students posed a threat to the quality of teaching and learning of science disciplines. Students were deprived of learning opportunities and access to resources. They lost the benefits of hands-on practical training and peer learning. Lecturers could not close the feedback loop on tests and assignments.

> P12: "… the biggest down side was the fact that we couldn't get [them] onto campus. That was like traumatic….. We saw that as the biggest hindrance to completing what we wanted to complete." and "The honours projects suffered the most, it was converted to a literature review and a research proposal, so that was a problem."

Theme 2: Effective Communication Is of Paramount Importance

Lecturers were frustrated by inadequate communication from the university, which increased their sense of insecurity. However, they realised the importance of effective communication with students and used a plethora of social media tools for communication and teaching. Lecturers had to find innovative ways to deal with floods of email messages from students.

> P11: We felt uninformed. Everything was a secret and we didn't know why the campus was closed and it was very, very frustrating.
>
> P12: What worked, I think is social media. I give them the gold medal…… All of a sudden everybody was on WhatsApp and that worked very well. People were really helping each other and talking to each other.
>
> P6: …they could write emails and we responded, sometimes we had hundreds of the same question and you would then upload the answer onto clickUP [LMS] for them.

Theme 3: Affordances and Challenges Associated with Teaching Technologies

The exceptional circumstances provided an impetus for experimentation with a range of teaching technologies, which presented a steep learning curve for many. Lecturers realised that they had to be sensitive to the data demands of virtual materials and guard against information overload, especially for poorly resourced and weaker students.

> P9: We would make videos of between 5 and 10 min and then we would upload this onto YouTube and embed that into clickUP [LMS]. So uhm the problems that we encountered there was that we had a number of software issues.

> P4: the information coming to the students from such a variety of such different directions, we have textbooks, we have lectures, we have online lectures, we have lecture slides, we have tutorial assignments, and for students to, especially the weaker students, to suddenly integrate all that information and to make it usable in an exam proves very difficult...... So what we had to do is to look at our content very clearly and be very aware of what is core, what would be difficult for the students, what is nice-to-have and to re-assess how we want to disseminate that information...

Theme 4: Concerns About Changed Assessment Practices

Online formative assessment worked well, but lecturers (and students) had limited experience with alternative modes of assessment that could replace traditional summative assessments. Concerns were raised about fairness and the validity of assessment that had to be changed without warning.

> P12: What we definitely will take on board is online tutorials. I love the fact that they work together. It doesn't count that much to the final mark so it doesn't matter if somebody is copying someone else's. They work together and they spoke to each other.

> P11: What went quite badly was assessment. Okay, we tried an exam assignment uhm and we had a single assessment....The problem with the single assessment is also they have no sort of yardstick for how they are actually going to be assessed.

Theme 5: Students Are Weak on Academic Self-regulation in Online Environment

Lecturers realised that students needed assistance towards self-regulation, i.e. they needed encouragement, guidance and pacing to keep up with the work.

> P11: What we learn from the students: they don't self-motivate, we know that, okay? They have a nasty habit of ...sort of they need to be egged on a little bit. We tend to assume that university students will operate themselves, but they don't.

> P7: It is however not enough to make these resources available to students, so we guided them through the process by... with a detailed study plan, as well as biweekly emails explaining what the focus of the day is and an assessment the following day to assess their level of understanding.

> P10: ... the lecturer as a metronome. Setting the pace explicitly. Telling them what to do, day by day. Guiding them, specifying what they should do.

Theme 6: Staff Resilience and Ethics of Care

Lecturers were traumatised due to sustained uncertainty, the constant need to adapt, high demands on their time and exhaustion. They nevertheless committed themselves to the academic project and to their students, often at cost to themselves. Many of them reported that they gained from the experience in terms of mastery of new teaching strategies and personal growth.

P1: So, although we felt a bit defeated and overwhelmed with all the additional admin that was on our desks ….. we still believed that we had a passion for our students and for our profession and to kinda get them to follow their careers in whichever way. So we had to step up our game, we had to be a bit more innovative and creative….

P11: We found it a very tough time, but I think there is also a potential to learn a lot of new skills that will move us away from traditional lecture strategies and something new and dynamic.

P12: I think we all showed a lot of restraint and resilience, patience and perseverance…

Discussion

The University of Pretoria was able to complete the 2016 academic year with a good level of confidence that rigour and fairness were served despite serious challenges. The enablers that made this possible were at least twofold: firstly, an infrastructure and an ecosystem that supported teaching innovation, and secondly, the resilience, commitment and ethics of care demonstrated by lecturers and support staff. The institutional contribution was to ensure that all students had the means and the opportunity to access e-learning platforms. Lecturers responded by implementing new ways of communication, teaching and assessment. These reflections by lecturers from a wide range of science disciplines set the scene for the study on chemistry students' experiences that will be reported next.

Part 2: Student Learning Preferences Versus Performance in Chemistry

The sudden shift to the OL environment disrupted traditional systems for the support and regulation of student learning, such as regular lectures and sit-down summative assessments. Under abnormal circumstances, self-paced learning combined with formative assessment becomes critically important to achieve the learning objectives in most disciplines, particularly in chemistry. The second part of the chapter focuses on tutorials offered in online (OL) or face-to-face (F2F) modality in an introductory organic chemistry course presented in the fourth quarter of the first year. In the absence of lectures and laboratory learning, this component of the course became the primary vehicle through which teaching and learning could be achieved. At the time of disruptions in September 2016, we were busy with a research project on student preferences for OL versus F2F learning in this course and the relationship of preference to performance. At the time, we had already collected data by means of a pre-course e-learning survey on access to the OL environment and learning preferences. The disruptions derailed the project as planned, but it created a once-off opportunity to explore student views after they had experienced almost fully F2F delivery as well as almost fully OL learning for the remaining six weeks of the course. Ethical clearance for the study was obtained from the ethics committee of the University of Pretoria (EC 160720-057).

CMY127 is a first-year, second-semester course consisting of physical and organic chemistry. It is a large enrolment gateway course ($N = 1386$) with pass rates ranging from 60 to 65% for the past four years. Students attend four 50 min lectures per week and one 3 h laboratory session fortnightly. Learning is supported by online homework (OWL for physical chemistry and Connect® for organic chemistry). At the time of campus closure, we were in the process of offering the first F2F tutorials in first-year chemistry for the organic component. This meant that some students did not get the chance to participate before the opportunity was lost. After campus closure, access to campus became very problematic and only *ca.* 200 students could attend weekly F2F tutorial sessions offered for students in this course. The rest of the cohort was fully dependent on OL learning.

We set out to answer the following research questions:

RQ1. How does students' learning preference relate to their performance in organic chemistry?
RQ2. What were the affordances and hindrances of F2F and OL learning as reported by students?
RQ3. Which students were challenged most by campus closure and for what reasons?

Data Collection and Analysis

The explanatory type of mixed methods design (Creswell & Clark, 2011) was used to answer these questions. Research questions one and three required a quantitative analysis of performance data and survey data collected prior to the disruptions; the second question was answered based on a qualitative analysis of survey data collected after the students had written their final examination.

The pre-course survey was designed, piloted and refined before implementation in 2016; the post-survey was constructed just in time for administration after the final CMY 127 examination. The pre-survey was completed by 1166 respondents (17 items on access and learning preferences, 84% response rate) and the post-survey by 565 respondents (21 items on experience, 41% response rate). The seven items in the pre-survey that were coded for online learning preference are provided in Appendix. Records of repeating students were removed, and the samples were stratified based on prior performance in Grade 12 physical sciences (PS), which was shown to be a good predictor for performance in first-year chemistry (Potgieter, Ackermann, & Fletcher, 2010) and biology (Kritzinger, Lemmens, & Potgieter, 2018). Three subgroups were formed, which were labelled *at-risk* (AR, achieved below 70% for Gr 12 PS), *murky middle* (MM, achieved 70–79% for PS in Grade 12) and *likely-to-pass* (LTP, achieved at least 80% for Gr 12 PS). The term "murky middle" reflects the fact that their outcomes were uncertain, with some students passing and others failing the course. The cleaned stratified sample consisted of 968 student records for

Table 6.1 Cross-tabulation of students according to their prior performance, online preference and performance in CMY 127 (average %)

	At-risk (AR)		Murky middle (MM)		Likely-to-pass (LTP)		All	
Online preference	*n*	Average %	*n*	Average %	*N*	Average %	*n*	Average %
High	142	45.9	102	53.0	55	60.6	299	51.0
Medium	141	48.2	89	54.0	59	62.6	289	53.1
Low	162	48.9	114	55.9	104	65.7	380	55.4
Total	445		305		218		968	

the pre-survey (AR 445, MM 305, LTP 218) and 435 records for the post-survey (AR 197, MM 136, LTP 97; 5 with missing Gr 12 data).

The open responses collected in the post-survey (435 responses to Q12) were coded inductively; more than one code may be assigned to the same response, which generated a total number of 622 assigned codes. Similar to the study presented in the first section, individual open coding was done by three researchers, followed by comparison and discussions, in order to reach an acceptable level of agreement on a set of seventeen themes that represent the data.

Findings

The pre-survey data on access indicated that only 10.6% of students did not have a suitable device and 14% did not have Internet access where they lived during term. These students would make use of free Internet access on campus under normal circumstances; however, after the campus was closed, they had to find another source for OL learning, for example to make use of the off-campus computer laboratories.

Learning Preference

RQ1: How does students' learning preference associate with performance in organic chemistry?

Responses to seven items in the pre-survey were categorised and scored for high, medium and low preference for OL learning, to calculate an OL preference total score and assign each student to one of the three subgroups. The cross-tabulation of students based on their OL preference, prospects for success and their performance in the CMY 127 chemistry course is reported in Table 6.1.

The results presented in Table 6.1 indicate that AR and MM students were divided almost equally between categories for OL preference, but this was not the case for the LTP, where low OL preference is dominant. In each of the performance bands, there was a negative correlation between preference for OL learning and performance in the CMY 127 course. Furthermore, students with the lowest preference for OL learning fared better when closure resulted in a largely OL learning environment. This result was contrary to our expectations; we anticipated that students would perform better if their learning preference is met.

OL preference by itself is a statistically significant predictor for success as measured by the CMY 127 examination mark ($p = 0.0001$). However, when controlling for prior performance, its influence is negated ($p = 0.11$), while the two explanatory variables for prior performance (G12 PS and CMY 117) are both highly significant ($p < 0.0001$).

Affordances and Hindrances of F2F and OL Learning

RQ2: What were the affordances and hindrances associated with each type of learning opportunity (F2F and OL) as reported by students?

The second research question probes the perceptions of students on the utility value of F2F versus OL tutorials after they experienced the abrupt change from almost fully F2F delivery to almost fully OL learning for the last six weeks of the course. Two items in the post-survey were analysed:

Q11: Which was more helpful Connect or Class Tutorials as presented during this semester?
Q12: Explain your preference for Connect or Class Tutorials. (Open response)

The majority of students (56.8%) found class tutorials to be more helpful despite the fact that there were few opportunities for this to happen. Student preference for OL Connect assignments was 39.1%, with 4.1% missing data for this item. Students who preferred F2F motivated their preference in terms of the affordances of the modality (F2F+) or the hindrances of the other (OL−) or both. Similarly, students who found OL more helpful explained their choice in terms of perceived value of OL learning (OL+) or dissatisfaction with F2F (F2F−) or both. The coding of responses for these two broad groupings generated sub-themes that were grouped into themes with common content. Only those themes that represent at least 5% of the total number of assigned codes are included in Table 6.2.

Students appreciated the opportunity to ask questions (F1) and get stepwise explanations (F2), from an individual (F3). They found that the level of questions was appropriate and a good match to that of examinations (F4) in contrast to the online questions which were not (F6).

Student preference for class tutorial sessions is strongly motivated by the perceived affordances of contact teaching and learning, rather than negative experiences and perceptions of OL learning. The fact that the data were collected after students had to resort to almost fully OL learning is meaningful. We interpret the absence of serious complaints about OL learning to indicate that the majority of students (56.8%) experienced OL learning as an acceptable alternative, but not their first choice. Of

Table 6.2 Preference for class tutorials motivated

Themes on perceived value of class tutorials based (F2F+) [323 of 374 codes for F2F preference, 86.4%]	% of codes for F2F preference
F1. Student-directed learning. Class tutorials enable the student to direct his/her own learning and to improve understanding. It provides opportunities to ask questions, to learn from mistakes and to get feedback on level of performance	23.8
F2. Satisfaction with format. Class tutorials are preferred because they are paper-based, stepwise solutions are provided, there is opportunity for interactive learning, and models are used. It is easy to get assistance, they are offered at no additional cost, and attendance is compulsory	18.7
F3. Interaction with a person (human being) has positive value. Students prefer explanations offered by a person, the individualised help that they receive and the opportunity for one-on-one interaction	17.1
F4. Satisfaction with experience. Class tutorials are useful as examination preparation because theoretical concepts are explained, the exercises are at an appropriate level, and there is opportunity for peer learning	15.2
F5. Satisfaction with assistance provided by tutors or lecturer. Students prefer tutor assistance to technology, because they are accessible, helpful and knowledgeable, they persist and they focus on appropriate content	11.5
Themes on dissatisfaction with online tutorials (OL-) (51 of 374 codes for OL preference, 13.6%)	
F6. Dissatisfaction with the quality of the learning experience offered by Connect, a web-based assignment and assessment platform. Students found the level of the questions to be inappropriate, explanations were lacking, there was no opportunity to ask questions, and it was confusing at times	5.3
F7–F9. Themes with prevalence less than 5%	

Table 6.3 Preference for online tutorials motivated

Themes on perceived value of OL (OL+) [159 of 248 codes for OL preference, 64.1%]	% of codes for OL preference
O1. OL matched how they learn best. Students preferred to work on their own, to set their own pace. OL provided an opportunity to practise and was suitable for examination preparation	15.3
O2. Satisfaction with learning experience. Students appreciated the feedback provided and the stepwise solutions for worked problems. The level was appropriate and they acquired technical skills	12.5
O3. Satisfaction with the OL platform. Students appreciated the opportunity to have multiple attempts to solve problems, the integration with the textbook, a wide range of questions, and hints, solutions or memoranda that were provided	12.5
O4. Convenience. Students appreciated the convenience of own time, own place and the accessibility of the OL platform	11.7
O5. Student-directed learning. Students appreciated the opportunity to figure it out by themselves, to access information to solve the problems and to learn from their mistakes	9.7
Themes on dissatisfaction with class tutorials (F2F-) (89 of 248 codes for OL preference, 35.9%)	
O6. Dissatisfaction with tutors. Students complained that the tutor–student ratio of tutors was inadequate, tutors did not explain well, they made false assumptions (about prior knowledge or preparation), or they gave incorrect guidance	17.7
O7. Dissatisfaction with class organisation and format. The most prevalent complaint was that memoranda were not provided. Students also disliked the lack of structure and the noisy class environment	6.9

note is the fact that only 3% of students made comments about data cost and other technical challenges experienced within the OL environment (F9) as a reason for preferring class tutorials.

Analysis of the explanations of students that found OL learning more helpful (39.1%) generated five themes on the affordances of OL learning and three themes on the perceived hindrances of the class tutorials with prevalence above 5% (Table 6.3). Dissatisfaction with class tutorials contributed more substantially to the motivations that students provided for this preference (36% as compared to 14% in the case of F2F tutorials).

Students preferred to work at their own pace (O1) and appreciated the value of immediate feedback (O2) for self-monitoring and deeper learning (O5). Preference for OL was also strongly motivated by dissatisfaction with the class tutorials due to the high student: tutor ratio and confusion caused by tutors giving different answers (O6) as well as the unavailability of worked out answers (O7).

To summarise, the perceived affordances of class tutorials were closely related to their format; they provided opportunities for interaction with people (lecturers, tutors and peers), enabled self-directed learning and provided learning opportunities at an appropriate level for examination preparation. This experience is hampered by a high student-to-staff ratio, tutors that are not knowledgeable and lack of feedback. The perceived affordances of OL learning were also closely related to its format; students could work at their own pace, appreciated immediate feedback and the scaffolding of learning, and the convenience and accessibility of the OL platform. Some hindrances were noted, such as exercises that did not match the level required for the course, the lack of opportunity to ask questions, data costs and challenges with the use of chemistry software.

Effect of Campus Closure on Vulnerable Students

RQ3. Which students were challenged most by campus closure and for what reasons?

The decision to close the campus for student access at a critical time in the semester was not taken lightly, but it raised concerns that students that were academically most vulnerable may have suffered the most under these exceptional circumstances. It was therefore important to try to determine which students were most seriously affected by the situation. The most pertinent question was one of access to devices and access to the Internet. Our results have shown that most students had access to at least one suitable device, but that 14% of students did not have Internet access where they were living during term. However, the university addressed this problem by setting up off-campus computer laboratories where students could access all learning material posted on clickUP (LMS) and could communicate electronically with their lecturers.

There is another more subtle question to answer, namely whether students with a high preference for OL learning, but who are metacognitively unaware and poorly self-regulated, may embrace OL learning without realising the risks involved. This concern is based on our findings that 142 at-risk students with high preference for OL learning achieved the lowest marks for the course (Table 6.1). It also resonates with the findings of a recent meta-analysis of more than 500,000 courses taken by over 40,000 community and technical college students in Washington State, USA, in 2004 (Xu & Jaggars, 2014). The authors compared students' ability to persist and earn strong grades in OL courses relative to their ability to do so in F2F courses. While all types of students in the study suffered decrements in performance in OL courses, some struggled more than others to adapt: males, younger students, black students and students with lower grade point averages.

We analysed the responses of 57 of the 142 at-risk students who had a high preference for OL learning that completed the post-survey, specifically their answers to questions 11 and 12. Five students were unable to attend any class tutorials and were therefore unable to make an informed judgement. Based on their experience, the remaining 52 students were split almost equally between finding class tutorials or OL assignments more helpful, 25 versus 27 students, respectively. We interpret this finding to indicate that learning preference should not be the deciding factor for the choice of learning modality. When left to their own devices, students may choose a less than optimal modality based on incorrect perceptions or convenience. This is especially important for weaker students or students who have little experience of OL learning. Self-regulation in the OL environment presents a formidable challenge—one that only experienced students manage well. We can conclude that the sudden switch to fully OL learning at the University of Pretoria most challenged those students who were academically weaker, especially those who were unaware of the challenges associated with such a learning environment.

Conclusions and Implications for Teaching and Learning

The experience described in this paper generated insights that are relevant to the design of blended/hybrid strategies for teaching and learning in general and for chemistry in particular. To highlight a few: lecturers and students must be accustomed to working online some of the time as part of a normal, twenty-first century approach to teaching and learning. However, blended learning, to be done well and to promote student engagement and success, must be designed on sound online teaching principles. While it is generally accepted that tertiary students will take responsibility for their own learning, our study indicated that this cannot be taken for granted, especially if normal practice is seriously disturbed. In order to ensure that students stay on task in OL learning, lecturers should set the pace explicitly and provide incentives, markers and deadlines, as well as feedback mechanisms for monitoring purposes.

Our results indicate that the F2F and OL learning environments both make a valuable contribution to teaching and learning. Students were clearly satisfied with the affordances of both modalities. This is an interesting finding; lecturers also realised that although OL teaching enabled them to complete the semester, it was not effective on its own, especially for subjects such as chemistry where F2F laboratory learning forms an integral part of the curriculum. Blended learning courses will cater better for the range of student needs and preferences than a course with only one modality. Furthermore, lecturers could consider designing flipped laboratories as research has shown that flip teaching is not only limited to classroom settings (Teo, Tan, Yan, Teo, & Yeo, 2014). The latter could be beneficial specifically when universities are forced to shut down as a result of unforeseen circumstances. F2F tutorials can be improved by ensuring that sessions are orderly and that tutors are well trained, knowledgeable and accessible. Attention should also be paid to the provision of feedback in contact sessions so that students can better monitor their own learning.

Finally, the unusual circumstances brought about by student disruptions in 2016 generated a renewed appreciation for the value of well-designed blended courses and an awareness of the disproportionate vulnerability of at-risk students in the online environment.

Acknowledgements We acknowledge an allocation from a DHET Teaching Development Grant for a SOTL project.

Appendix: Seven Items from the Pre-course Survey Coded for Online Preference

Responses to seven items in the pre-survey were coded as indicated below. The sample was divided into three groups of roughly equal size (high, medium and low) based on their scores for OL preference.

Question 6 (Multiple Choice):
If you have used Connect platform previously, rate how well it assisted you to master the course(s) (e.g. MLB or PHY).

	Not used connect before	1. Not at all helpful	2	3	4	5	6	7	8	9	10. Very helpful
Code	0	1	1	1	1	2	2	2	3	3	3

Question 7: (Multiple Choice):
If you have used Learnsmart previously, rate how well it assisted you to master the course.

	Not used learnsmart before	1. Not at all helpful	2	3	4	5	6	7	8	9	10. Very helpful
Code	0	1	1	1	1	2	2	2	3	3	3

Question 9 (Multiple Choice) (coded in conjunction with Question 17):
If there is something you do not understand in class, how do you prefer to get additional information?

	Online	Textbook	Lecturer	Tutor	Classmate	Another person	Unanswered
Codes for OL preference	3	1[a]	1	1	1	1	0

[a]Coded as 3 if students selected e-book in Question 17

Question 13 (Multiple Choice):
When studying for a semester test, I will rely on online material for my learning:

	Never	Rarely	Usually	Always	Unanswered
Codes for OL preference	1	2	3	4	0

Question 14 (Multiple Choice):
Which type of compulsory tutorial do you think would help you understand the course better?

	Class tutorials	Online tutorials
Codes for OL preference	1	3

Question 15 (Multiple Choice):
Which type of compulsory tutorial do you think you would enjoy more?

	Class tutorials	Online tutorials
Codes for OL preference	1	4

Question 17 (Multiple Choice):
Do you prefer to read from a (i) paper-based book; (ii) e-book?

References

Acton, T., Scott, M., & Hill, S. (2005). E-education-keys to success for organizations. In 18th Bled eConference eIntegration in action. Sloveni, 6–8 June. https://domino.fov.uni-mb.si/proceedings.nsf/0/b2f6ff8014823eabc12570140049 16dc/$file/12acton.pdf. Accessed 3 August, 2017.

Adams Becker, S., Cummins, M., Davis, A., Freeman, A., Hall Giesinger, C., & Ananthanarayanan, V. (2017). NMC horizon report: 2017 Higher Education Edition. https://www.sconul.ac.uk/sites/default/files/documents/2017-nmc-horizon-report-he-EN.pdf. Accessed 7 August, 2017.

Akkoyunlu, B., & Soylu, M. Y. (2008). A study of student perceptions in a blended learning environment based on different learning styles. *Educational Technology and Society, 11*(1), 183–193.

Alammary, A., Sheard, J., & Carbone, A. (2014). Blended learning in higher education: Three different design approaches. *Australasian Journal of Educational Technology, 30*(4), 440–453.

Beyer, A. M. (2011). Improving student presentations: Pecha Kucha and just plain PowerPoint. *Teaching of Psychology, 38*(2), 122–126.

Brooks, R. (2015). Politics and protest—students rise up worldwide. *University World News, 13,*413. http://www.universityworldnews.com. Accessed 25 September, 2017.

Brown, M. G. (2016). Blended instructional practice: A review of the empirical literature on instructors' adoption and use of online tools in face to face teaching. *Internet and Higher Education, 31,* 1–10.

Chandra, V., & Fisher, D. L. (2009). Students' perceptions of a blended web-based learning environment. *Learning Environment Research, 12,* 31–44.
Chen, C. C., & Jones, K. T. (2007). Blended learning vs traditional classroom settings: assessing effectiveness and student perceptions in an MBA accounting course. *Journal of Educators online, 4*(1), 1–15.
Christiansen, M. A. (2014). Inverted teaching: applying a new pedagogy to a university organic chemistry class. *Journal of Chemistry Education, 91,* 1845–1850.
Creswell, J. W. (2009). *Research design: Qualitative, quantitative and mixed methods approaches.* Thousand Oaks: Sage Publications.
Creswell, J. W., & Clark, V. L. (2011). *Designing and conducting mixed methods research.* Thousand Oaks: Sage Publications.
Dziuban, C., Graham, C. R., Moskal, P. D., Norberg, A., & Sicilia, N. (2018). Blended learning: the new normal and emerging technologies. *International Journal of Educational Technology in Higher Education, 15*(3), 1–16.
Flyn, A. B. (2015). Structure and evaluation of flipped chemistry courses: organic and spectroscopy, large and small, first to third year, English and French. *Chemistry Education and Research Practice, 16,* 198–211.
Graham, C. R., Woodfield, W. J., & Harrison, B. (2013). A framework for institutional adoption and implementation of blended learning in higher education. *Internet and Higher Education, 18,* 4–14.
Hill, T., Chidambaram, L., & Summers, J. (2013). A field experiment in blended learning. Performance effects of supplementing traditional classroom experience with web-based virtual learning environment. In *13th AMCIS Proceedings 15–17 August 2013.* https://aisel.aisnet.org/amcis2013/ISEducation/GeneralPresentations/5. Accessed 1 August, 2017.
Ibrahim, Y. (2010). Between revolution and defeat: Student protest cycles and networks. *Sociology Compass, 4*(7), 495–504.
Isilow, H. (2016, October 5). South Africa university shuts for week after protests. http://aa.com.tr/en/africa/south-africa-university-shuts-for-week-after-protests/658825. Accessed 14 June, 2017.
Kritzinger, A., Lemmens, J.-C., & Potgieter, M. (2018). Learning strategies for first-year biology: Toward moving the "Murky Middle". *CBE–Life Sciences Education, 17*: ar42, 1–13.
Lewis, S. (2016, June 8). Papua New Guinea moves to end student protests after violence. *Time*, Available at: www.time.com. Accessed 25 September, 2017.
Lincoln, Y. S., & Guba, E. G. (1989). *Fourth generation evaluation.* Newbury Park: Sage.
Marin, M. (2012, April 22). Student protestors in Quebec are safeguarding the future. University World News. www.universityworldnews.com. Accessed 25 September, 2017.
Ossiannilson, E. (2017). Blended learning: State of the Nation. https://icde.memberclicks.net/assets/RESOURCES/BlendedLearningICDEInsightPaper2017compressed.pdf. Accessed 6 September, 2018.
Potgieter, M., Ackermann, M., & Fletcher, L. (2010). Inaccuracy of self-evaluation as additional variable for prediction of students at risk of failing first-year chemistry. *Chemistry Education Research and Practice, 11,* 17–24.
Reid, S. A. (2016). A flipped classroom redesign in general chemistry. *Chemistry Education Research and Practice, 17,* 914–922.
Renner, D., Laumer, S., & Weitzel, T. (2014). Effectiveness and efficiency of blended learning: A literature review. In *20th Americas conference on information systems proceedings 7–9 August 2014.* https://aisel.aisnet.org/cgi/viewcontent.cgi?article=1286&context=amcis2014. Accessed 11 August, 2017.
Smith, J. D. (2013). Student attitudes toward flipping the general chemistry classroom. *Chemistry Education Research and Practice, 14,* 607–614.
Tekane, R. Louw, I., & Potgieter, M. (2018). #Fees must fall: Science Teaching during Student Unrest. *Alternation, 25,* 161–180.
Teo, T. W., Tan, K. C. D., Yan, Y. K., Teo, Y. C., & Yeo, L. W. (2014). How flip teaching supports undergraduate chemistry laboratory. *Chemistry Education Research and Practice, 15,* 550–567.

Twigg, C. A. (2003). Improving learning and reducing costs: New models for online learning. *Educause Review, 38*(5), 28–38.

Xu, D., & Jaggars, S. S. (2014). Performance gaps between online and face-to-face courses: Differences across types of students and academic subject areas. *The Journal of Higher Education, 85*(5), 633–659.

Yeung, K., & O'Malley, P. J. (2014). Making the flip work: barriers to and implementation strategies for introducing flipped teaching methods into traditional higher education courses. *New Directions, 10,* 59–63.

Chapter 7
Bridging the Gap Between Philosophy of Science and Student Mechanistic Reasoning

Nicole Graulich and Ira Caspari

Abstract Reasoning about mechanisms is central to understanding relevant components and cause–effect relations of phenomena in science. Explaining mechanisms is, thus, an essential practice in natural sciences and STEM disciplines. Accounts from philosophy of science provide insight into the nature of mechanisms and mechanistic reasoning and how knowledge about mechanisms is gathered and evaluated. Mechanistic reasoning comprises reasoning about the entities and activities involved in a process and the way these entities and activities are organized. These theoretical considerations combined with an educational perspective on students' learning provide a lens for the methodology of educational studies, e.g. for the analysis of students' productive resources or for curriculum changes. This chapter outlines how accounts from philosophy of science have been used to inform research on student learning from primary school children's descriptions of physical phenomena to multi-level reasoning processes in undergraduate biology classes. Depending on the nature of the discipline and the educational objective of the study, different aspects of what constitutes mechanistic reasoning from philosophy of science have been adapted. Specifically, we illustrate how accounts from philosophy of science helped us characterize students' reasoning processes about organic reaction mechanisms.

Introduction

The ability to construct mechanistic explanations by asking *how* and *why* a phenomenon occurs, rather than just describing the *what* of a mechanistic process, is essential for successful scientific reasoning (Abrams, Southerland, & Cummins, 2001; DiSessa, 1983; Duncan, 2007; Hammer & Berland, 2012; Cooper, 2015). Answering those questions requires reasoning about different components of a system, i.e. entities, their properties, and the activities they undergo when producing

N. Graulich (✉) · I. Caspari
Justus-Liebig University Gießen, Institute of Chemistry Education, Gießen, Germany
e-mail: Nicole.Graulich@didaktik.chemie.uni-giessen.de

I. Caspari
e-mail: Ira.caspari@didaktik.chemie.uni-giessen.de

M. Schultz et al. (eds.), *Research and Practice in Chemistry Education*,
https://doi.org/10.1007/978-981-13-6998-8_7

the phenomenon (Machamer, Darden, & Craver, 2000). Providing such mechanistic explanations for scientific phenomena is considered to be a constituent part of scientific enquiry (Russ, Scherr, Hammer, & Mikeska, 2008; Koslowski & Masnick, 2002), in addition to the ability to build arguments (Brown, Furtak, Timms, Nagashima, & Wilson, 2010; Driver, Newton, & Osborne, 2000) or to use analogical reasoning (Kaminski, Sloutsky & Heckler, 2013; Duit, 1991).

However, the vagueness of what constitutes a mechanism and what accounts for mechanistic reasoning makes it difficult for researchers and teachers to recognize students' reasoning as "mechanistic" and to provide appropriate instructions to promote their efforts. In search for defining what constitutes mechanistic reasoning, theoretical accounts, for instance, of Bechtel and Abrahamsen (2005), Glennan (1996, 2002), and particularly of Machamer et al. (2000) described how scientific knowledge about mechanisms is gathered through the explicit search, analysis, and explanation of constituent parts of mechanisms. Machamer et al.'s (2000) philosophical work gained attention as they were purposefully observing and describing working scientists' strategies and characterized the process as a search for essential components of mechanisms. While trying to explain mechanisms in their discipline, working scientists aim at identifying entities and activities and defining the relationships between them including their spatial and temporal organization to fully account for phenomena from starting to end points. This search offers answers to typical "How does it work?" questions, which scientists ask about phenomena.

Machamer et al. (2000) demonstrated this central goal of searching, describing, and explaining mechanisms by referring to the mechanism of the protein synthesis and multi-level mechanisms in neuroscience. They described mechanisms as:

> composed of both entities (with their properties) and activities. Activities are the producers of change. Entities are the things that engage in activities. [...] The organization of these entities and activities determines the ways in which they produce the phenomenon. (Machamer et al., 2000, p. 3)

Reasoning about phenomena is, thus, mechanistic in nature if it specifies the entities and activities, as well as their properties and organization, and the cause–effect relationships that account for the sequential stages of the phenomena (Illari & Williamson, 2012; Tabery, 2004). The overall mechanism is the change of entities from setup to termination conditions. Fundamental to this description of mechanisms is the dualism of entities and activities, i.e. the fact that entities with certain properties engage in activities, which change the entities and their properties and enable them to engage in different activities. Accordingly, describing mechanisms is more than reasoning about starting and end points of a mechanism statically; it also requires considering how activities of various entities bring about the change in a dynamic way. Hence, mechanistic reasoning can be considered as the act of accounting for a phenomenon by (1) identifying the entities of the mechanism and their properties, (2) identifying the activities that the entities engage in, (3) identifying the temporal and spatial organization of entities and activities, and (4) connecting those components to account for the dynamic sequence of mechanistic steps. Machamer et al. (2000) described the function of explaining mechanisms by stating that "an explanation renders a phenomenon intelligible [...] and shows *how possibly, how plausibly* or *how actually* things work" (Machamer et al., 2000, p. 21). Reasoning about mechanisms, thus, provides a rationale for the "productive continuity without gaps from the set up to termination conditions" (Machamer et al., 2000, p. 3). One can only fully account for a phenomenon if one outlines how causes bring about effects (Schauble, 1996) by explaining how properties can be used to causally connect reasoning about entities and activities because properties of entities enable the activities they undergo and activities change the properties of entities. Darden (2002), Darden and Craver (2002) later used the term "chaining" to describe a reasoning strategy that scientists use to reason about unknown parts of the mechanism based on known parts of the mechanism. This reasoning strategy uses the causal connection between entities, their properties, and activities and demonstrates that these components can logically be chained together to account for the phenomenon.

Describing the underlying cause–effect relations between entities and activities does often involve reasoning about implicit, i.e. non-observable, properties of entities. This is often challenging for learners in chemistry (cf. Graulich, 2015). When learners rationalize mechanisms, they tend to use domain-general heuristics, teleological, or mechanical descriptive approaches (Talanquer, 2018).

Application of a Philosophical Framework in Science Education

In search for frameworks to support the analysis of student reasoning and understanding of complex systems, the philosophical accounts on the nature of mechanisms in science from Machamer et al. (2000) and others (Glennan, 2002; Illari & Williamson, 2012) have been adopted in research studies in various science disciplines. As these philosophical accounts have not been described with an educational purpose, each study used their own adaptation with regard to the type of learner, e.g. primary school children (Russ, Scherr, Hammer, & Mikeska, 2008) or pre-university students (van Mil, Postma, Boerwinkel, Klaasen, & Waarlo, 2016), and the type of mechanism and discipline under investigation, i.e. the physical phenomenon of falling objects (Russ et al. 2008) or processes at the molecular level of a cell (van Mil et al. 2013; van Mil et al. 2016). Each of the aforementioned examples as well as other subsequent studies (Moreira, Marzabal, & Talanquer, 2018; Southard, Wince, Meddleton, & Bolger, 2016) used ideas from philosophy of science to purposefully inform educational research to analyse or scaffold students' reasoning. In the following sections, we illustrate how two of the first approaches used theoretical considerations in philosophy of science to support the analysis of student mechanistic reasoning.

Students' Reasoning About Mechanisms in Physics

One of the first adaptations of Machamer et al.'s (2000) philosophical account has been conducted in physics education by Russ et al. (2008). They aimed at analysing narratives of primary school children's descriptions of a physics phenomenon, i.e. falling objects, and derived coding categories from Machamer et al.'s (2000) philosophical account to provide a means to recognize incipient stages of mechanistic reasoning in students' discourse. Seven coding categories have been defined, (1) describing the target phenomenon, (2) identifying setup conditions, (3) identifying entities, (4) identifying activities, (5) identifying properties of entities, (6) identifying organization of entities, and (7) chaining. They further added two additional categories, (8) analogy and (9) animated models, to capture primary school children's use of analogies and non-verbal gestures or movements that they may use for describing the phenomenon. There is no predefined hierarchy between these codes that makes an explanation mechanistic or not; however, some of the codes are interdependent, e.g. the description of organization of entities should logically follow the identification of entities.

Russ et al. (2008) illustrated that their practical adaptation of Machamer et al.'s (2000) approach can be used in a discourse analysis of a first-grade classroom discussion to describe the occurrence of students' mechanistic reasoning. Their focus on identifying the "substantial pieces" of mechanistic reasoning makes it easier for educators to identify mechanistic reasoning in their students' discourse. Using the coding categories as a lens allows to identify students' progress and to help students fill in gaps in their mechanistic explanations.

Others used these coding categories to qualitatively compare opportunities for mechanistic reasoning in two instructional settings, such as open and guided enquiry laboratory activities (Wulf et al. 2013). The coding revealed that students' written notes in an open enquiry showed more incidences of *identifying setup conditions, identifying activities, identifying properties of entities*, and *chaining* than in the more guided curriculum.

Russ et al.'s (2008) and Wulf et al.'s (2013) work focused on identifying components of mechanistic reasoning and makes, thus, students' use of these components during enquiry or description of phenomena accessible. A current research approach in general chemistry builds upon this approach and combines it with a focus on students' causal models while explaining phenomena (Moreira et al., 2018).

Thinking About Multiple Levels of Mechanisms in Biology Education

Machamer et al.'s (2000) framework originally used examples from molecular biology, which makes their philosophical account appealing to biology educators. Van Mil et al. (2013) followed Machamer et al.'s (2000) aspect of hierarchically organized entities and activities to support molecular mechanistic reasoning in biology education. Each phenomenon in biology can be explained at different organizational levels, e.g. describing the change of entities at the higher level of cells and organisms or on a molecular level, e.g. from properties of functional modules to functional group interactions. In biology classes, however, phenomena are often discussed at a cellular level (e.g. the function of DNA for cell duplication), rather than at the molecular level or levels in between the molecular and the cellular level, i.e. understanding the functional groups of the DNA and how this knowledge can be helpful to understand macromolecular mechanisms and relationships. Thus, biology students struggle to connect cellular, macromolecular, and molecular levels. Adopting the idea of reasoning about entities, activities, and their organization helped van Mil et al. (2013) conceptualize mechanistic reasoning as the search for the respective entities and activities at multiple levels. Their approach highlights that reasoning about the components of a mechanism connects systemic properties at the cellular level with activities of molecules at a lower level. The entities and their activities identified at higher levels can be used to hypothesize other entities, activities, and organization at lower levels, and reasoning back and forth between levels can be used to fully account for the mechanism.

To address biology students' difficulties to reason about processes at different levels, van Mil et al. (2016) designed an instructional approach that builds upon their model of molecular mechanistic reasoning. The instruction scaffolded students' analysis of each level—displayed as an animation, cartoon, or drawing—and asked the students to connect the displayed entities or activities to the other levels. Connecting questions focused on helping students interpret the animations and drawings of cellular processes and connect these to the molecular and subcellular levels, e.g. What is represented? Why does it matter? How is it related to the other model?

Russ et al.'s (2008) and van Mil et al.'s (2013) work are the first examples of how theoretical considerations in philosophy of science about mechanistic reasoning provide a starting point for analysing and scaffolding student mechanistic reasoning. Each approach made its changes and adaptations to the original description and made it applicable in an educational context.

Using Aspects of Philosophy of Science to Analyse Students' Reasoning in Organic Chemistry

The term "mechanism" is ubiquitous in organic chemistry classes, and the ability to propose plausible reaction mechanisms for organic transformations is one of the central goals of teaching and learning. Despite the omnipresence of the term, insights from philosophy of science or philosophy of organic chemistry have not been used as a theoretical basis to guide studies on student mechanistic reasoning previous to our account. The following two studies illustrate how philosophical accounts can be beneficial for analysing student mechanistic reasoning in organic chemistry.

Connecting Components of a Mechanism—Chaining

To account for the productive continuity from the starting to the end point of a mechanism in the event that only parts of the mechanism are known, one can reason about subsequent or former stages in a mechanism by using the entities, their properties, and the activities that are already known (Craver & Darden, 2013). This act of *chaining* rationalizes how components are connected and why they bring about a change. Chaining in mechanistic reasoning answers questions regarding the connection of entities and activities in various stages of the observed phenomenon (Darden, 2002; Darden & Craver, 2002): What property of an entity gave rise to this activity?, What activity will this entity engage in?, or What activity gave rise to this entity with these properties?

When one predicts a mechanism, given the starting materials and products, one may start the reasoning process with the product and rationalize back to its formation, or one may consider the starting materials and their properties and hypothesize an activity they may undergo. Craver and Darden (2013) described these two directions of reasoning as forward and backward chaining:

> In forward chaining one uses the early stages of a mechanism to reason about the type of entities and activities that are likely to be found in later stages. In backward chaining one reasons from the entities and activities in later stages in a mechanism to find entities and activities appearing earlier. (Craver & Darden, 2013, p. 77)

In both cases, chaining can be either directed towards the activity (activity-directed) or towards the entity (entity-directed) of a later or a former stage of the mechanism. As such, four types of chaining can be described (Caspari, Weinrich, Sevian & Graulich, 2018b). Activity-directed forward chaining means deriving an unknown activity that a given entity can undergo. Entity-directed forward chaining considers an entity and the activity it undergoes to infer the resulting entity. Proposing a typical mechanistic step of an organic reaction can be divided in an entity- and an activity-directed forward chaining step (Fig. 7.1).

Backward chaining, divided into activity-directed and entity-directed backward chaining, means reasoning from one component of the mechanism to a prior activity or prior entity. This reasoning pattern is opposite to the direction of the reaction mechanism.

In organic chemistry, activity-directed backward chaining can be necessary in the event that (1) forward chaining does not help determine which of two or more alternative activities to represent, and (2) if the activity under consideration has some interdependences with subsequent activities that lead to the energetically favoured pathway (Caspari et al. 2018b). The following example illustrates a situation where activity-directed backward chaining is necessary and appropriate to use (Fig. 7.2). During the first step of an ester hydrolysis, a lone pair of the nucleophilic hydroxide is attracted by the partially positive carbon of the ester functionality. A tetrahedral intermediate is formed. Forward chaining from this intermediate results in two competing pathways: one is the reverse reaction restoring the ester and the nucleophile (Fig. 7.2. step a) and the other is the departure of the alkoxide, resulting in a car-

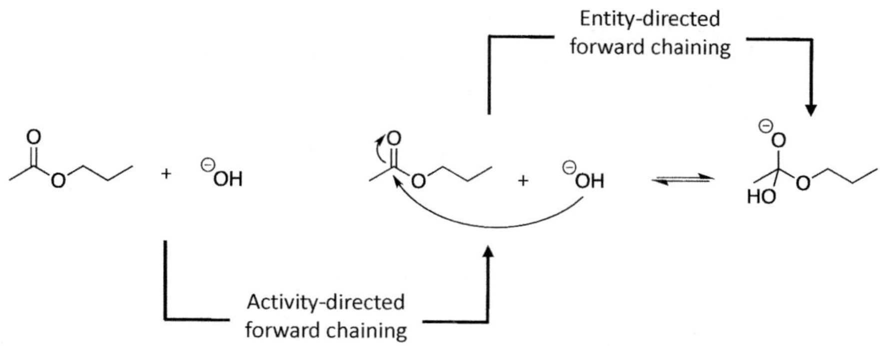

Fig. 7.1 Two types of forward chaining in the reasoning process about a mechanistic step of an organic reaction

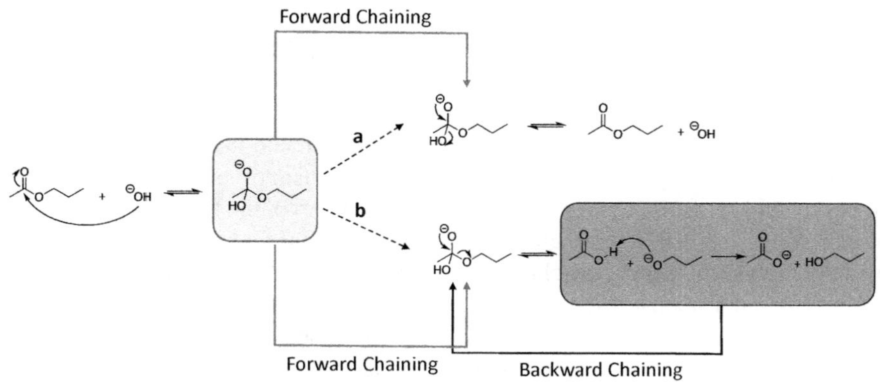

Fig. 7.2 Illustration of forward and backward chaining between competing pathways

boxylic acid (Fig. 7.2. step b). Both pathways are reasonable and have comparable activation barriers. Thus, forward chaining does not help to determine which of the two activities to represent. Only in the case of the second pathway (Fig. 7.2. step b), a resonance-stabilized carboxylate is formed in the subsequent acid–base reaction, leading to an energy minimum and making it essentially impossible for the reverse reaction to occur. Hence, there is an interdependence of this later mechanistic step with the departure of the alkoxide, namely that the equilibrium of the departure step is constantly shifted towards the alkoxide. Based on the alternative outcomes of forward chaining and this interdependence, backward chaining can be employed in this context to make the decision that the departure of the alkoxide is the activity, i.e. the electron movement, to represent (Fig. 7.2. step b).

Students use these chaining types when reasoning about organic reaction mechanisms, but they are not always aware of which context is appropriate to employ a specific chaining type. When students propose a mechanism, they tend to map the starting material onto the product, trying to get to the product of the mechanism. This is largely known as the “It gets me to the product strategy” (Bhattacharyya & Bodner, 2005). The direction of this problem-solving strategy is—at least in parts—reversed relative to the direction of the reaction mechanism because information about the end stage of the mechanism is taken to make predictions about earlier stages. While backward chaining can be necessary in some cases, using it as a general problem-solving approach without reflecting the underlying chemical causality can lead to application of the reasoning strategy in inappropriate contexts and to chemically unsound mechanistic predictions.

Considering backward chaining as a typical reasoning strategy when reflecting on mechanistic steps provides a lens to understand how information about mechanistic steps is collected and conjectured in organic chemistry and in which cases backward chaining is a valid strategy. Investigating how students apply backward chaining gives valuable information on how students can be supported building causal mechanistic explanations. Thus, we were interested in whether students employ backward chaining causally or non-causally and whether they do so in appropriate or inappropriate contexts (Caspari et al. 2018b). Students’ backward chaining was further compared to their professor’s expectations about the students’ reasoning. For this objective, we performed a secondary data analysis using data previously collected in the USA (Sevian et al. 2015) that focused on students’ and their professor’s explanations while solving typical organic reaction mechanism problems, in our case the Jones oxidation, the PCC mechanism, and two examples of the Baeyer–Villiger Oxidation. Semi-structured, think-aloud interviews had been conducted with 20 students and the professor teaching the class after the first and the third of three examinations of the Organic Chemistry II course. Each interview included two mechanism problems. The transcripts of these interviews were analysed qualitatively based on the theoretical considerations about chaining derived from philosophy of science.

All of the students used activity-directed backward chaining to reason about former mechanistic steps. Although half of the participants used this chaining type in mechanistic contexts, in which it was appropriate, i.e. alternative pathways of the mechanistic step were plausible and interdependencies with later mechanistic steps were given, none of the students mentioned alternatives or included energetic considerations that would causally explain the interdependencies. The other half of the students did also not include these considerations and employed backward chaining in an inappropriate context.

Comparing students’ approaches to their professor’s expectations revealed that he expected the students to engage in backward chaining in appropriate contexts. He mentioned plausible alternative pathways of the respective mechanisms, but did not discuss the energetic consideration to decide between pathways. As such, his expectations and the reasoning strategies that the students employed were both non-causal in nature.

Through the adaptation of chaining as a lens for analysis, this study revealed that analysing students’ approaches to typical mechanism problems and their professor’s expectations helps understand the origin of students’ non-causal backward chaining that is not sufficient to differentiate between appropriate and inappropriate contexts (Caspari et al. 2018b). Furthermore, it shaped the understanding of how reasoning about mechanisms is communicated in teaching. When teaching multi-step mechanisms in organic chemistry classes, such as the Baeyer–Villiger Oxidation, often only the lowest energy path is displayed and information about how to decide between alternative pathways is not provided.

Structural and Energetic Accounts

Information about the role of energetic considerations in reasoning about organic reaction mechanisms can as well be drawn from philosophy of organic chemistry, a field of philosophy of science advanced by Goodwin (2003, 2008). He describes what type of domain-specific questions an organic chemist would ask when investigating mechanisms. Those questions are all questions about transformations and ask for the rates of reactions (e.g. which reaction proceeds faster?) or ask for preferred pathways (e.g. which pathway A or B is preferred in a reaction?). He further exemplifies what sort of considerations is required to provide answers to typical questions of the discipline. Goodwin (2003) describes that such an answer in organic chemistry can be divided into a structural account, which is a claim about structure, and an energetic account, i.e. a claim about the energy of a reaction. To build a structural account about the difference in rate of two mechanistic steps, one needs to consider information implicitly encoded in the structural representations, i.e. information about the electronic structure of entities. For instance, when deciding which halide is the best leaving group in a substitution reaction, a typical structural account would be the following statement: iodide compared to chloride is the largest halide, and negative charge that accumulates on the leaving group in the transition state is accommodated best (Fig. 7.3). We further transferred Goodwin's theoretical considerations in philosophy of organic chemistry to organic chemistry education and specified aspects of Goodwin's work to make it applicable for analysis of student data (Caspari, Kranz, & Graulich, 2018a). When analysing the given example, one focuses on an explicit structural difference represented in the structural formulas (e.g. iodide (A) versus chloride (B), Fig. 7.3), and based on this difference, one infers an implicit structural difference (e.g. larger size of the iodide). This implicit property of the entity is then related to a change in the process (e.g. the formation of the negative charge) by stating that the negative charge forming during the departure step is better accommodated in case of the larger iodide. The idea of two distinct accounts translates into the idealized reasoning structure represented in Fig. 7.3.

As such, Goodwin's philosophical account provided a starting point and a guidance to theoretically describe how relations between properties of entities and changes—changes of properties of entities that occur during activities—should be constructed to build a structural account of an answer to a mechanistic comparison of two cases. An energetic account of the answer can then be derived from the structural account, e.g. stating the relative activation energies of the two processes.

This general reasoning scheme offers a lens to capture students' ability to build relations between an explicit difference and the change that occurs in two mechanistic

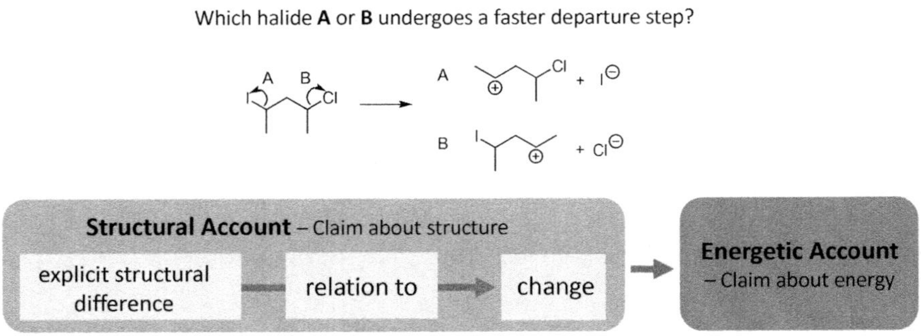

Fig. 7.3 Illustration of the reasoning structure of an answer to a typical mechanistic question in organic chemistry

steps. In a qualitative think-aloud study, we asked 20 chemistry and food chemistry majors of an undergraduate Organic Chemistry II course to predict the relative activation energy of two contrasted leaving group departure steps. The transcripts of these interviews were then used to characterize the complexity of relations that students constructed while solving these case comparisons (Caspari et al. 2018a). Students' relations between explicit differences and the changes in the process differed in how much implicit information the students considered between the starting point, i.e. the explicit difference, and the end point of the relation, i.e. the change. In an unfamiliar context, students, who were able to verbalize more implicit information, were more often successful in predicting the effect of a difference on a change correctly.

We further added a second layer of analysis by explaining how the aspect of "change" can be approached when thinking about mechanisms. The description of "change" as the end point of a relation can be approached statically or dynamically. This dualism of statics and dynamics when reasoning about mechanisms is also demonstrated in the philosophy of science literature (Bechtel & Abrahamsen, 2005; Glennan, 2002; Illari & Williamson, 2012; Machamer et al., 2000). If one compares the potential energy of two halides and two carbocations relative to their haloalkanes, one can approach change statically, for example, by only considering the formal charges present in the products but not the process that gave rise to them. However, to compare the activation energy of two reactions, one needs to approach change dynamically, i.e. one needs to consider the actual process and the formation of new properties in the process, e.g. the formation of positive and negative charges (Goodwin, 2003). We found that students approached change statically in about half of the cases although a claim about activation energy, and thus, a dynamic approach to change was required. Hence, the dualism of statics and dynamics, prevalent in philosophy of science, helped us compare what a certain type of questions requires and how students actually approach the question.

A purposeful combination of insights from philosophy of science and philosophy of organic chemistry allowed us to describe productive resources in students' mechanistic reasoning and to uncover future areas of research

(Caspari et al., 2018a). Especially, the connection of static structures and dynamic reasoning about processes and energetic aspects were difficult for students. Future initiatives should focus on supporting students to become aware of their productive resources, which they already use in their reasoning. Scaffolding can build on this and support students to use explicit properties as the starting point of considering implicit properties and static perspectives on change as the starting point for dynamic perspectives on change.

Conclusion

Using insights from philosophy of science to inform educational research is one of various ways to approach student mechanistic reasoning. The research efforts presented herein have shown why it may be valuable for educators to adopt philosophical considerations to guide or to inspire their educational studies. The approaches outlined used philosophical considerations either as an accessible tool to analyse students' discourse, as an aid to restructure and connect different levels of biological models, or as a guidance to describe domain-specific reasoning structures in organic chemistry. The adaptation of these philosophical theories in educational studies with different disciplinary backgrounds, thus, bridges the gap in a twofold way, on the one hand between philosophy of science and student mechanistic reasoning and on the other hand between science disciplines. These research approaches in science education invite the education community to reflect on commonalities and differences between types of reasoning in science disciplines, offering a route to a general conceptualization of mechanistic reasoning in science education.

References

Abrams, E., Southerland, S., & Cummins, C. (2001). The how's and why's of biological change: How learners neglect physical mechanisms in their search for meaning. *International Journal of Science Education, 23*(12), 1271–1281. https://doi.org/10.1080/09500690110038558.

Bechtel, W., & Abrahamsen, A. (2005). Explanation: A mechanist alternative. *Studies in History and Philosophy of Science Part C: Studies in History and Philosophy of Biological and Biomedical Sciences, 36*(2), 421–441. https://doi.org/10.1016/j.shpsc.2005.03.010.

Bhattacharyya, G., & Bodner, G. M. (2005). "It gets me to the product": How students propose organic mechanisms. *Journal of Chemical Education, 82*(9), 1402–1407. https://doi.org/10.1021/ed082p1402.

Brown, N. J. S., Furtak, E. M., Timms, M., Nagashima, S. O., & Wilson, M. (2010). The evidence-based reasoning framework: Assessing scientific reasoning. *Educational Assessment, 15*(3–4), 123–141. https://doi.org/10.1080/10627197.2010.530551.

Caspari, I., Kranz, D., & Graulich, N. (2018a). Resolving the complexity of organic chemistry students' reasoning through the lens of a mechanistic framework. *Chemistry Education Research and Practice, 19*(4), 1117–1140. https://doi.org/10.1039/c8rp00131f.

Caspari, I., Weinrich, M., Sevian, H., & Graulich, N. (2018b). This mechanistic step is "productive": Organic chemistry students' backward-oriented reasoning. *Chemistry Education Research and Practice, 19*(1), 42–59. https://doi.org/10.1039/C7RP00124J.

Cooper, M. M. (2015). Why ask why? *Journal of Chemical Education, 92*(8), 1273–1279. https://doi.org/10.1021/acs.jchemed.5b00203.

Craver, C. F., & Darden, L. (2013). *In search of mechanisms: Discoveries across the life sciences*. Chicago, IL: University of Chicago Press.

Darden, L. (2002). Strategies for discovering mechanisms: Schema instantiation, modular subassembly, forward/backward chaining. *Philosophy of Science, 69*(S3), 354–365. https://doi.org/10.1086/341858.

Darden, L., & Craver, C. (2002). Strategies in the interfield discovery of the mechanism of protein synthesis. *Studies in History and Philosophy of Science Part C: Studies in History and Philosophy of Biological and Biomedical Sciences, 33*(1), 1–28.

DiSessa, A. A. (1983). Phenomenology and the evolution of intuition. In D. Gentner & A. L. Stevens (Eds.), *Mental models* (pp. 15–34). Hillsdale, NJ: Lawrence Erlbaum Associates Inc.

Driver, R., Newton, P., & Osborne, J. (2000). Establishing the norms of scientific argumentation in classrooms. *Science Education, 84*(3), 287–312. https://doi.org/10.1002/(Sici)1098-237x(200005)84:3%3c287:Aid-Sce1%3e3.0.Co;2-A.

Duit, R. (1991). On the role of analogies and metaphors in learning science. *Science Education, 75*(6), 649–672. https://doi.org/10.1002/sce.3730750606.

Duncan, R. G. (2007). The role of domain-specific knowledge in generative reasoning about complicated multileveled phenomena. *Cognition and Instruction, 25*(4), 271–336. https://doi.org/10.1080/07370000701632355.

Glennan, S. (1996). Mechanisms and the nature of causation. *Erkenntnis, 44*(1), 49–71.

Glennan, S. (2002). Rethinking mechanistic explanation. *Philosophy of Science, 69*(3), S342–S353. https://doi.org/10.1086/341857.

Goodwin, W. (2003). Explanation in organic chemistry. *Annals of the New York Academy of Science, 988*, 141–153.

Goodwin, W. (2008). Structural formulas and explanation in organic chemistry. *Foundations of Chemistry, 10*(2), 117–127. https://doi.org/10.1007/s10698-007-9033-2.

Graulich, N. (2015). The tip of the iceberg in organic chemistry classes: How do students deal with the invisible? *Chemistry Education Research and Practice, 16*, 9–21. https://doi.org/10.1039/c4rp00165f.

Hammer, D., & Berland, L. K. (2012). Framing for scientific argumentation. *Journal of Research in Science Teaching, 49*(1), 68–94. https://doi.org/10.1002/tea.20446.

Illari, P. M., & Williamson, J. (2012). What is a mechanism? Thinking about mechanisms across the sciences. *European Journal for Philosophy of Science, 2*(1), 119–135. https://doi.org/10.1007/s13194-011-0038-2.

Kaminski, J. A., Sloutsky, V. M., & Heckler, A. F. (2013). The cost of concreteness: The effect of nonessential information on analogical transfer. *Journal of Experimental Psychology: Applied, 19*(1), 14–29. https://doi.org/10.1037/a0031931.

Koslowski, B., & Masnick, A. (2002). The development of causal reasoning. *Blackwell Handbook of Childhood Cognitive Development*, 257–281.

Machamer, P., Darden, L., & Craver, C. F. (2000). Thinking about mechanisms. *Philosophy of Science, 67*(1), 1–25. https://doi.org/10.1086/392759.

Moreira, P., Marzabal, A., & Talanquer, V. (2018). Using a mechanistic framework to characterise chemistry students' reasoning in written explanations. *Chemistry Education Research and Practice*. https://doi.org/10.1039/c8rp00159f.

Russ, R. S., Scherr, R. E., Hammer, D., & Mikeska, J. (2008). Recognizing mechanistic reasoning in student scientific inquiry: A framework for discourse analysis developed from philosophy of science. *Science Education, 92*(3), 499–525. https://doi.org/10.1002/sce.20264.

Schauble, L. (1996). The development of scientific reasoning in knowledge-rich contexts. *Developmental Psychology, 32*(1), 102–119. https://doi.org/10.1037/0012-1649.32.1.102.

Sevian, H., Bernholt, S., Szteinberg, G. A., Auguste, S., & Perez, L. C. (2015). Use of representation mapping to capture abstraction in problem solving in different courses in chemistry. *Chemistry Education Research and Practice, 16*(3), 429–446. https://doi.org/10.1039/c5rp00030k.

Southard, K., Wince, T., Meddleton, S., & Bolger, M. S. (2016). Features of knowledge building in biology: Understanding undergraduate students' ideas about molecular mechanisms. *CBE-Life Sciences Education, 15*(1), 1–16. https://doi.org/10.1187/cbe.15-05-0114.

Tabery, J. G. (2004). Synthesizing activities and interactions in the concept of a mechanism. *Philosophy of Science, 71*(1), 1–15. https://doi.org/10.1086/381409.

Talanquer, V. (2018). Exploring mechanistic reasoning in chemistry. In J. Yeo, T. W. Teo, & K.-S. Tang (Eds.), *Science education research and practice in Asia-Pacific and beyond* (pp. 39–52). Singapore: Springer Singapore. https://doi.org/10.1007/978-981-10-5149-4_3.

van Mil, M. H., Boerwinkel, D. J., & Waarlo, A. J. (2013). Modelling molecular mechanisms: A framework of scientific reasoning to construct molecular-level explanations for cellular behaviour. *Science & Education, 22*(1), 93–118. https://doi.org/10.1007/s11191-011-9379-7.

van Mil, M. H. W., Postma, P. A., Boerwinkel, D. J., Klaassen, K., & Waarlo, A. J. (2016). Molecular mechanistic reasoning: Toward bridging the gap between the molecular and cellular levels in life science education. *Science Education, 100*(3), 517–585. https://doi.org/10.1002/sce.21215.

Wulf, R., Hinko, K., & Finkelstein, N. (2013). *Comparing mechanistic reasoning in open and guided inquiry physics activities*. Paper presented at the Physics Education Research Conference, Portland, OR. July 2013.

Chapter 8
Assessing for Chemical Thinking

Vicente Talanquer

Abstract Current educational reform efforts emphasize the importance of developing students' abilities to engage in a variety of science practices, including creating and using models, generating arguments, and building explanations. To achieve these goals, we need to change more than our instructional methods. It is also critical that we carefully reflect on how to modify our assessment tools to better elicit the types of understandings that we value. In this chapter, I describe a strategy we are using to develop formal formative assessments and summative assessment in a general chemistry course for science and engineering majors in the USA. These assessments seek to evaluate the extent to which students can integrate central ideas, crosscutting ways of reasoning, and core practices in making sense of relevant phenomena and in finding reasonable solutions to real problems. The benefits and challenges of our approach to the assessment of student understanding are highlighted in this chapter.

Introduction

The goals for science education in general and chemistry education in particular, as described in recent educational policy documents across the world (EC, 2015; NRC, 2011, 2013) are quite ambitious. We would like our students to design investigations of relevant phenomena, collect and analyse data, build and use models to make sense of these data, engage in argumentation based on evidence, and generate explanations using central ideas in a discipline. Meeting these learning objectives demands major changes in how we traditionally approach chemistry instruction, how we conceptualize curricula, and how we assess student understanding.

Seeking to better align our educational goals and practices with those advocated by researchers in science and chemistry education, for the past 10 years we have worked to transform the one-year general chemistry course for science and engineering majors at our university (Talanquer & Pollard, 2010, 2017). Our goal has been to develop a learning environment in which students actively engage in construct-

V. Talanquer (✉)
Department of Chemistry and Biochemistry, University of Arizona, Tucson, USA
e-mail: vicente@u.arizona.edu

M. Schultz et al. (eds.), *Research and Practice in Chemistry Education*,
https://doi.org/10.1007/978-981-13-6998-8_8

ing and applying chemical thinking to analyse, discuss, reflect upon, and propose reasonable explanations and solutions to relevant problems and phenomena. As part of our approach to educational reform, we have also changed the ways in which we design both formal formative assessments and summative assessments of student understanding. In this contribution, I summarize core elements of our approach to the assessment of student learning and discuss the benefits and challenges that have arisen when working with various instructors, dozens of graduate teaching assistants, and thousands of students every semester.

Our Educational Goals

To better understand our views of assessment, it is critical to first describe the core educational goals of our redesigned general chemistry course. Traditional approaches to the teaching of introductory chemistry often emphasize the acquisition of fundamental chemical knowledge in diverse areas, such as stoichiometry, atomic and molecular structure, thermodynamics, and kinetics. In contrast, in our "Chemical Thinking" curriculum, we seek to help students understand and be able to apply a core set of central ideas, ways of reasoning, and practices in chemistry (Talanquer & Pollard, 2010). The central ideas include: substance characterization, structure determination, property prediction, reaction analysis, reaction control, and sustainable action (Talanquer, 2016). These ideas enable ways of knowing, thinking, and acting that are useful in answering six essential questions in our discipline: How do we identify chemical substances? How do we predict the properties of substances? Why do chemical substances change? How do chemical substances change? How do we control the changes that substances undergo? And, what is the impact of chemical action?

Using an educational model and philosophy analogous to the one advanced by the Next Generation Science Standards (NGSS) in the USA (NRC, 2011, 2013), in our general chemistry course, we confront students with relevant phenomena and problems for which the explanation or solution demands the integration of the central ideas, crosscutting ways of reasoning, and core practices represented in Fig. 8.1. The integration of these three dimensions of understanding characterization what we define as chemical thinking (Sevian & Talanquer, 2014). To scaffold the development of students' ability to productively engage in this type of thinking, we carefully select the sequence of problems to be presented to students throughout the course, basing our choices on existing research on how student understanding of core concepts in chemistry typically progresses with education (Cooper, Underwood, Hilley, & Klymkowsky, 2012; Smith, Wiser, Anderson, & Krajcik, 2006; Talanquer, 2009, 2018a).

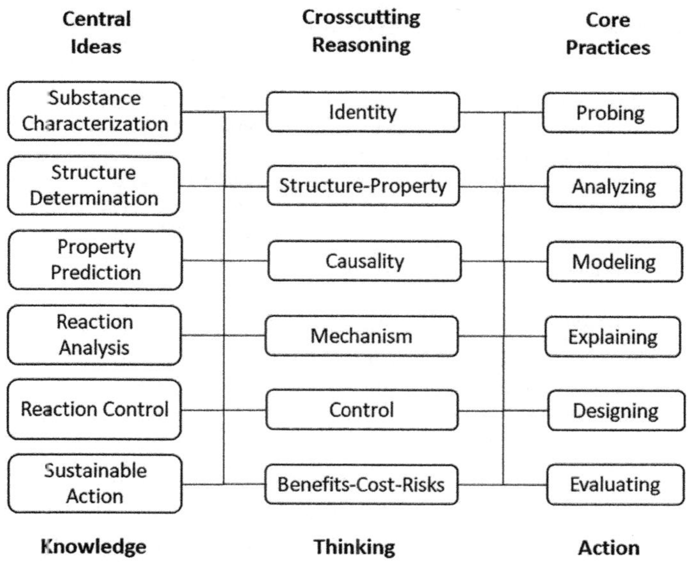

Fig. 8.1 Three dimensions of understanding that students are expected to integrate through their work in the Chemical Thinking curriculum

Our Summative Assessments

Given our educational goals, we have changed our approach to the assessment of student understanding. We now seek to develop assessment instruments that provide evidence of students' ability to integrate central ideas, crosscutting ways of reasoning, and core practices when confronted with phenomena or problems that are analogous to those analysed and discussed in class, but situated in different contexts. In the following paragraphs, I describe and discuss our approach to the construction of these formal formative and summative assessments using a concrete example to illustrate the different factors that we take into consideration in the development of our assessment tools.

The construction of an assessment begins with the identification of sets of assessment targets, each of them including a central idea of interest and associated crosscutting ways of reasoning and core practices. One of these sets is illustrated in Table 8.1 for a case involving the understanding of structure-property relationships in ionic compounds. Each set of assessment targets is then used to build an "integrated performance" that describes in general terms how students are expected to demonstrate their ability to integrate the idea, ways of reasoning, and practices that are highlighted. These integrated performances are equivalent to the performance expectations included in the NGSS for K-12 education in the USA (NRC, 2013), which make explicit the knowledge and thinking in action that students should proficiently demonstrate by the end of instruction.

Table 8.1 Example of assessment targets and associated integrated performance

Dimension	Assessment targets	Integrated performance
Idea	The physical properties of ionic substances depend on the charge and size of the ions that comprise them. These latter properties have periodic behaviours	Students analyse data about physical properties of ionic compounds and build models to explain observed patterns in terms of the electrostatic interactions between the ions that comprise the substances
Reasoning	Relate macroscopic properties of substances to the interaction and behaviour of particles at submicroscopic levels	
Practice	Build models based on data that explain and predict measured properties of systems and processes	

Once a core set of integrated performances has been defined, we focus our efforts on identifying a relevant context in which such understandings can be applied to analyse, comprehend, explain, or solve an issue of relevance. The scenario should be rich enough to facilitate the evaluation of students' integrated and transferable understandings using diverse types of questions and problems. However, the scenario should not be so complex that it cannot be meaningfully analysed by students at the introductory chemistry level. Periodical publications by the American Chemical Society (ACS) such as ChemMatters and Chemical and Engineering News (C&EN) are often great sources of ideas for potential scenarios.

With a scenario at hand, we look for information and data that can be used to develop the setting for students to analyse. Then, we work to specify the different practices students should engage in, the chemical knowledge they should demonstrate, and the types of reasoning they are expected to apply. This level of specification is illustrated in Table 8.2 for the assessment targets listed in Table 8.1. In this example, the scenario focuses on the analysis of the physical properties of ionic liquids being considered as potentially greener substitutes of volatile organic solvents commonly used in research laboratories and industrial settings.

The scenario in Table 8.2 creates diverse opportunities for exploring students' ability to integrate knowledge, thinking, and action. Students may be asked to, for example, analyse data, identify trends, construct and apply submicroscopic models of matter to build explanations, engage in structure-property reasoning, as well as evaluate the environmental and economic benefits and costs of modifying chemical practices. These opportunities are transformed into the different tasks that students will be asked to complete as part of the formal formative or summative assessment. Specific examples of potential assessment tasks for the scenario under considerations are also included in Table 8.2.

Examples of the types of assessments that we have generated using the approach outlined in previous paragraphs are freely available online at https://sites.google.com/site/chemicalthinking/tasks, and a section of the actual instrument generated using

Table 8.2 Assessment scenario and associated assessment tasks

Dimension	Scenario specifications	Assessment tasks
Setting	Ionic liquids are chemical compounds with relatively low melting points that show promise in many areas: battery electrolytes, green solvents. Their melting point depends on their chemical composition and structure	• Analyse the data provided to identify correlations between melting points and the composition and structure of targeted ionic compounds • Create visual representations of the different substances at the submicroscopic level showing relative values for the charge and size of the ions • Build and apply a model to explain the trends in the data based on the composition and structure of the ionic compounds • Evaluate potential environmental and economic benefits and costs of using ionic solvents instead of volatile organic compounds
Practices	Students will analyse experimental data about the melting points of different ionic liquids. They will propose a submicroscopic model that explains observed experimental patterns based on the charge and size of the ions present in each compound. Students will use this model to predict the relative melting points of other related substances	
Ideas	This task demands for students to understand how ion size and charge affect the magnitude of electrostatic interactions between ions, and how those properties vary for different types of atoms and ions depending on their composition and structure	
Reasoning	The task demands building rationales that connect interactions between ions at the submicroscopic level and measurable physical properties of chemical compounds at the macroscopic level	

the "ionic liquids" scenario is included in Fig. 8.2. These examples illustrate both the assessment targets that are valued within the Chemical Thinking curriculum, and our efforts to make visible to students how those targeted understandings are of central importance in explaining relevant phenomena or addressing important problems in modern societies.

Design and Implementation Benefits and Challenges

The development of the types of formative and summative assessments described in this chapter may be perceived as a daunting task. However, the work is greatly simplified once assessment targets are specified for core components of the course. It also helps to adopt a collaborative approach in the development of assessment instruments. At our institution, all general chemistry instructors work together in

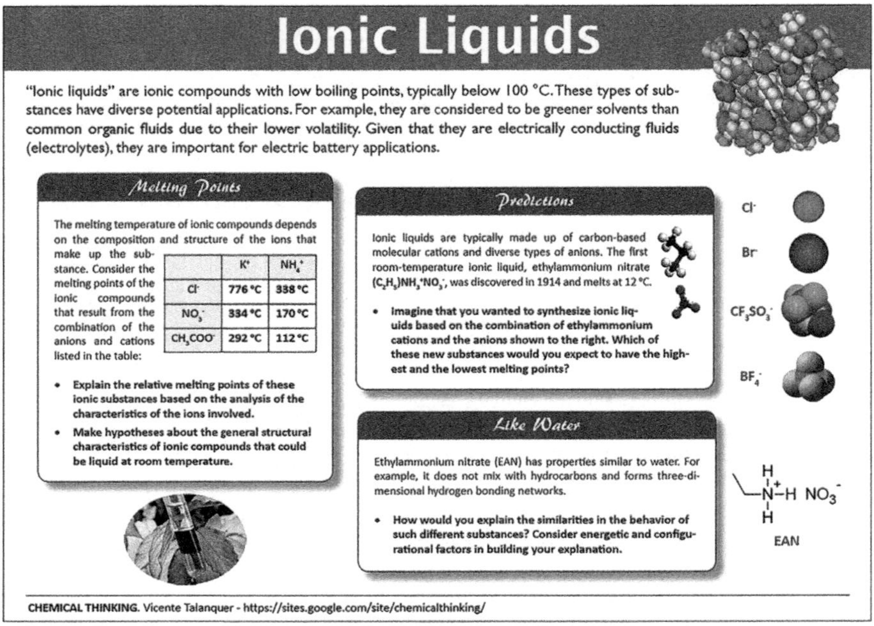

Ionic Liquids

"Ionic liquids" are ionic compounds with low boiling points, typically below 100 °C. These types of substances have diverse potential applications. For example, they are considered to be greener solvents than common organic fluids due to their lower volatility. Given that they are electrically conducting fluids (electrolytes), they are important for electric battery applications.

Melting Points

The melting temperature of ionic compounds depends on the composition and structure of the ions that make up the substance. Consider the melting points of the ionic compounds that result from the combination of the anions and cations listed in the table:

	K^+	NH_4^+
Cl^-	776 °C	338 °C
NO_3^-	334 °C	170 °C
CH_3COO^-	292 °C	112 °C

- **Explain the relative melting points of these ionic substances based on the analysis of the characteristics of the ions involved.**
- **Make hypotheses about the general structural characteristics of ionic compounds that could be liquid at room temperature.**

Predictions

Ionic liquids are typically made up of carbon-based molecular cations and diverse types of anions. The first room-temperature ionic liquid, ethylammonium nitrate $(C_2H_5)NH_3^+NO_3^-$, was discovered in 1914 and melts at 12 °C.

- **Imagine that you wanted to synthesize ionic liquids based on the combination of ethylammonium cations and the anions shown to the right. Which of these new substances would you expect to have the highest and the lowest melting points?**

Cl^-

Br^-

$CF_3SO_3^-$

BF_4^-

Like Water

Ethylammonium nitrate (EAN) has properties similar to water. For example, it does not mix with hydrocarbons and forms three-dimensional hydrogen bonding networks.

- **How would you explain the similarities in the behavior of such different substances? Consider energetic and configurational factors in building your explanation.**

NO_3^-

EAN

CHEMICAL THINKING. Vicente Talanquer - https://sites.google.com/site/chemicalthinking/

Fig. 8.2 Section of a formal formative assessment in the Chemical Thinking curriculum

the design of the summative assessments to be implemented every year, and these instruments get reused as formal formative assessments in subsequent semesters. Individual instructors volunteer to take the lead in identifying potential scenarios for a given assessment and proposing potential assessment tasks. The initial proposal is then enriched through several rounds of discussion involving all instructors. This approach has helped us clarify our assessment targets and develop a shared vision for what we want students to understand and how these understandings should be demonstrated.

Our work on the construction of scenario-based integrated assessments has led us to reflect on the types of reasoning that we would like our students to express when confronted with our assessment tasks. We noticed, for example, that we had to find a better balance between tasks that asked students to demonstrate their understanding of *what* was happening in a system or phenomenon versus tasks that required them to show understanding of *why* and *how* it was happening. We realized that we were investing considerable effort in assessing students' ability to interpret and create representations of different systems and phenomena at different levels, but there were fewer opportunities for students to use those representations in the construction of chemical rationales to support justifications (e.g. when making predictions), arguments (e.g. when making claims based on data provided), and explanations (e.g. when rationalizing a phenomenon based on chemical ideas).

As we have embarked in developing more assessment tasks that probe students' abilities to build interpretations, justifications, arguments, and explanations, we have become aware of the need to be more explicit about the types of chemical rationales that we are expecting students to construct (Talanquer, 2018b), as well as about the modes of reasoning we want them to apply (Sevian & Talanquer, 2014). Creating opportunities for students to integrate in productive manners the three dimensions of chemical understanding outlined in Fig. 8.1, and building assessment instruments that evaluate such a competence, demand careful analysis of the specific types of thinking we expect students to manifest.

Chemical scientists use their knowledge to build different types of rationales to justify, explain, or argue about the properties and behaviours of chemical substances. Consider, for example, the three different rationales included in Table 8.3 for why aspirin is classified as an acid. Each of these rationales may be an acceptable answer depending on the context in which the question is posed and on the types of understandings that students are expected to demonstrate. We thus have found it critical to more clearly define the nature of the chemical concepts, ideas, and models that we want students to invoke during a given task.

Chemical justification, arguments, and explanations can also be built at different levels of complexity in terms of the types and number of associations or causal relationships that are invoked, from simple correlations to linear causal relations to mechanistic models that take into account and integrate the effects of multiple properties and interactions. In this regard, we have also found important to explicitly press students to build their justifications, arguments, and explanations using causal mechanistic reasoning, rather than simple causal or relational reasoning (Sevian & Talanquer, 2014). To illustrate this point, consider the different explanations for why ethanol dissolves in water included in Table 8.4. At first glance, the different explanations in this table may be judged reasonable and appropriate depending on the context. However, we value the development of students' ability to build mechanistic explanations in which the properties and behaviours of the target system or phenomenon (e.g. ethanol dissolving in water) are constructed in reference to the properties, interactions, and organization of relevant components typically acting at a sublevel (e.g. molecular, atomic, electronic) (Talanquer, 2018c; Russ et al., 2008).

Making explicit the types of chemical rationales and modes of reasoning we want students to use in assessment tasks has led us to modify our teaching practices to foster the type of learning that we deem important. We have sought, for example, to more explicitly define the types of chemical mechanisms that we would like our students to

Table 8.3 Different types of chemical rationales

Chemical rationale	Example
Phenomenological	Aspirin is an acid because its aqueous solutions have pH < 7 (it increases [H+])
Mechanical	Aspirin is an acid because its molecules may transfer a proton when interacting with water molecules
Structural	Aspirin is an acid because its molecules have a carboxylic acid group and the loss of a proton creates an anion that is stabilized by resonance

Table 8.4 Different modes of reasoning

Modes of reasoning	Example
Relational	Ethanol is soluble in water because its molecules are polar
Simple causal	Ethanol dissolves in water due to dipole–dipole and H-bonding interactions between water and ethanol molecules
Mechanistic	Dipole–dipole and H-bonding interactions between water and molecules cause ethanol molecules to separate from each other. Ethanol molecules are then dispersed among the solvent molecules

meaningfully understand and be able to apply when confronted with different types of tasks in the general chemistry course. We have organized these mechanisms into five major groups related to (Talanquer, 2018c): (1) matter transformation, (2) energy transfer and transformation, (3) activation, (4) stabilization, and (5) equilibration, and we are working to consistently create opportunities for students to recognize when and how these different mechanisms can be productively deployed to build explanations or support arguments.

We have also recognized the need to better scaffold the development of students' ability to generate mechanistic explanations. We are, for example, exploring the effect of using reasoning scaffolds like the one depicted in Fig. 8.3. The goal is to guide students in the identification of the different elements that support the construction of mechanistic rationales, making explicit major reasoning elements such as: identifying relevant components in a system and in the associated system model; characterizing the properties of these components; recognizing interactions between components and the processes to which these interactions lead (both in terms of matter transformation and energy transfer and transformation); and integrating these different elements into a mechanistic story. We work with students to help them recognize the meaning and importance of the different parts of the scaffold, providing multiple opportunities for them to construct them in different contexts.

Clarifying our goals and making visible our expectations has been helpful not only to our students, but also to the dozens of teaching assistants (TAs) that support student learning in the classroom and the laboratory. These TAs are responsible for evaluating student work in summative assessments in the general chemistry course. One of the major challenges we have faced in the implementation of our assessment approach is the training of the TAs who are to judge the quality of students' answers to the assessment tasks that we pose. Most of these TAs have little teaching experience and hold rather traditional beliefs about teaching and learning. Given the nature of their own experiences as students in traditional chemistry courses, they often fail to see the value of pressing students for mechanistic explanations or demanding different types of chemical rationales. Making more explicit our learning goals and consistently using them in the design of instructional activities implemented in the classroom and the laboratory has been a productive strategy in the difficult task of changing TAs' teaching beliefs and reframing their expectations about student learning.

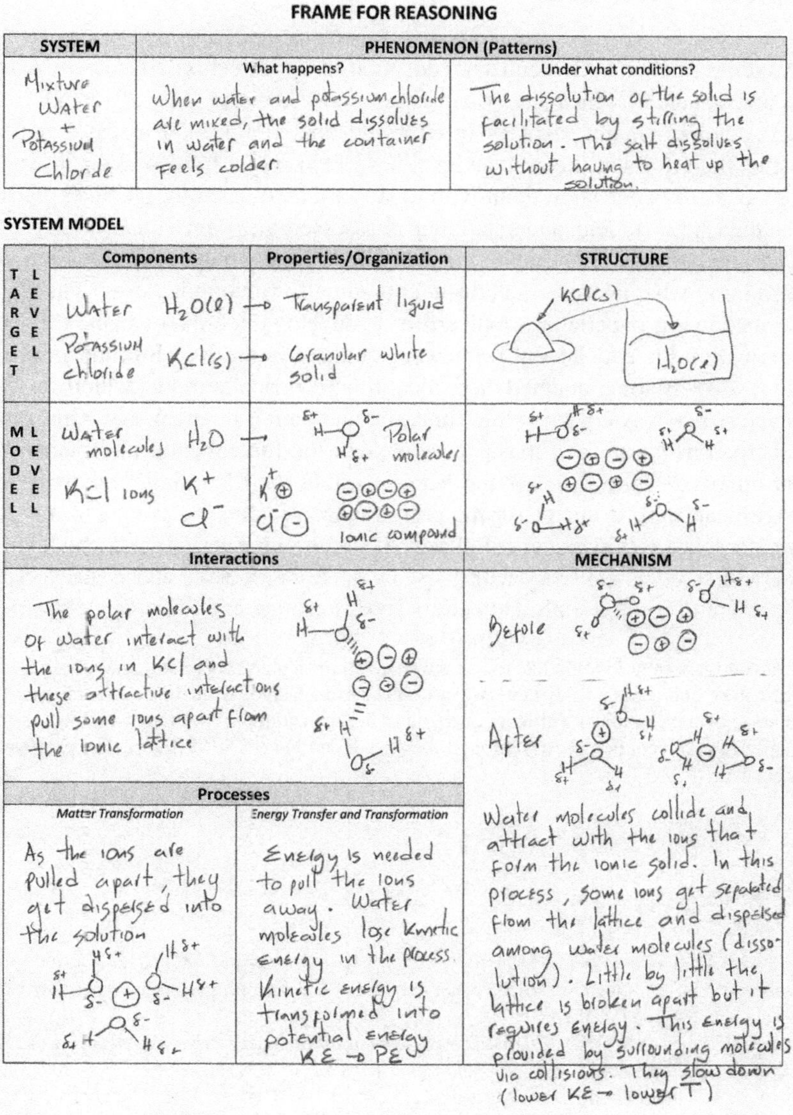

Fig. 8.3 Example of scaffolding framework used to guide student engagement in mechanistic reasoning

Concluding Remarks

Advancing science and chemistry education in the direction advocated by current education standards in many countries around the world (EC, 2015; NRC, 2011, 2013) demands major changes in curricula, instruction, and assessment. So far, most of the efforts at the college level have focused on transforming instructional approaches to better align them with evidence-based teaching practices. Initiatives to change how student understanding is assessed are rarer (Laverty et al., 2016; Underwood et al., 2018) despite the centrality of assessment in communicating and determining what instructors and students actually value in learning. The approach described in this paper aligns with efforts to develop three-dimensional (3D) assessments within the educational framework outlined in the NGSS in the USA (NRC, 2011, 2013). We have adapted these ideas to our curricular model which emphasizes characteristics ways of knowing, thinking, and acting in chemistry. This approach has helped us clarify and make more explicit the fundamental understandings we want our students to develop and better scaffold their learning. Our work has led to significant improvements in the performance of diverse students in our general chemistry classes (Talanquer & Pollard, 2017), although various challenges still need to be addressed as we work with large numbers of students and instructors with a wide range of backgrounds and beliefs about teaching and learning.

Acknowledgements I would like to acknowledge all the general chemistry instructors, laboratory coordinators, and graduate assistants who are and have been involved in the development and implementation of the Chemical Thinking curriculum at our institution. Support from the Association of American Universities (AAU) through their Undergraduate STEM Education Initiative is also appreciated.

References

Cooper, M. M., Underwood, S. M., Hilley, C. Z., & Klymkowsky, M. W. (2012). Development and assessment of a molecular structure and properties learning progression. *Journal of Chemical Education, 89,* 1351–1357.

European Commission (EC). (2015). *Science education for responsible citizenship*. Luxembourg: European Commission.

Laverty, J. T., Underwood, S. M., Matz, R. L., Posey, L. A., Carmel, J. H., Caballero, M. D., et al. (2016). Characterizing college science assessments: The three-dimensional learning assessment protocol. *PLoS ONE, 11*(9), e0162333.

National Research Council (NRC). (2011). *A framework for K-12 science education: Practices, crosscutting concepts, and core ideas*. Washington DC: The National Academies Press.

National Research Council (NRC). (2013). *The next generation science standards*. Washington DC: The National Academies Press.

Russ, R. S., Scherr, R. E., Hammer, D., & Mikeska, J. (2008). Recognizing mechanistic reasoning in student scientific inquiry: A framework for discourse analysis developed from philosophy of science. *Science Education, 92*(3), 499–525.

Sevian, H., & Talanquer, V. (2014). Rethinking chemistry: A learning progression on chemical thinking. *Chemistry Education Research and Practice, 15*(1), 10–23.

Smith, C., Wiser, M., Anderson, C., & Krajcik, J. (2006). Implications of research on children's learning for standards and assessment: A proposed learning progression for matter and atomic-molecular theory. *Measurement, 14*(1&2), 1–98.

Talanquer, V. (2009). On cognitive constraints and learning progressions: The case of structure of matter. *International Journal of Science Education, 31*(15), 2123–2136.

Talanquer, V. (2016). Central ideas in chemistry: An alternative perspective. *Journal of Chemical Education, 93*, 3–8.

Talanquer, V. (2018a). Progressions in reasoning about structure-property relationships. *Chemistry Education Research and Practice, 19*, 998–1009.

Talanquer, V. (2018b). Chemical rationales: Another triplet for chemical thinking. *International Journal of Science Education, 15*, 1874–1890.

Talanquer, V. (2018c). The importance of understanding fundamental chemical mechanisms. *Journal of Chemical Education, 95*(11), 1905–1911.

Talanquer, V., & Pollard, J. (2010). Let's teach how we think instead of what we know. *Chemistry Education Research and Practice, 11*(2), 74–83.

Talanquer, V., & Pollard, J. (2017). Reforming a large foundational course: Successes and challenges. *Journal of Chemical Education, 94*, 1844–1851.

Underwood, S. M., Posey, L. A., Herrington, D. G., Carmel, J. H., & Cooper, M. M. (2018). Adapting assessment tasks to support three-dimensional learning. *Journal of Chemical Education, 95*(2), 207–217.

Chapter 9
Assessment of Practical Chemistry in England: An Analysis of Scientific Methods Assessed in High-Stakes Examinations

Sibel Erduran, Alison Cullinane and Stephen John Wooding

Abstract "Practical science" has been widely used in the curriculum, exam board specifications and research traditions in science education for several decades in England. The terminology typically refers to a range of experimental and investigative activities conducted as part of science education in schools and colleges. According to major reviews of research literature, there is evidence that the assessment regime has had a major impact on practical work that teachers carry out. However, there is growing concern that the amount and quality of practical work carried out in schools suffer as a result of the impact of the high-stakes national tests. The chapter aims to investigate the underlying scientific methods that are promoted in the chemistry examination papers thus facilitating understanding of what is likely to be taught in chemistry lessons. In order to identify the types of scientific methods included in the chemistry examination papers, a framework was used focusing on four categories: manipulative hypothesis testing, non-manipulative hypothesis testing, manipulative parameter measurement and non-manipulative parameter measurement. The examination items from two examination papers of a leading examination board are classified according to these categories, and patterns on the marking are traced. The results indicate that for both papers, non-manipulative parameter measurement was the method assessed at a higher percentage. In both papers, manipulative hypothesis testing was the category with the lowest percentage of items or questions. Furthermore, the mark allocation was the highest in both papers in the non-manipulative parameter measurement category. The results indicate that there is consistency between the items allocated to each category of scientific methods, and the marks allocated to them, although in one paper there were more marks allocated to manipulative parameter measurement even though the relative frequency of this category was the lowest in the items .

S. Erduran (✉) · A. Cullinane
Department of Education, University of Oxford, 15 Norham Gardens, Oxford OX2 6PY, UK
e-mail: Sibel.Erduran@education.ox.ac.uk

A. Cullinane
e-mail: Alison.Cullinane@education.ox.ac.uk

S. J. Wooding
Centre for Education Research and Practice, AQA, Devas St., Manchester M15 6EX, UK
e-mail: swooding@aqa.org.uk

M. Schultz et al. (eds.), *Research and Practice in Chemistry Education*,
https://doi.org/10.1007/978-981-13-6998-8_9

This observation may reflect an assumption that manipulative parameter measurement is considered to be cognitively more demanding and thus deserving of more marks. Some implications for assessment of practical chemistry are discussed.

Introduction

Recent curriculum reform efforts about the inclusion of scientific practices in science teaching and learning have implied a closer investigation into what practical science gets assessed in high stakes examinations. For example, in the context of the recent developments with the Next Generation Science Standards in the USA, Reed, Brandriet and Holme (2017) investigated how science practices have been inherently incorporated into previously released American Chemical Society (ACS) examinations through analysis and classification of ACS general chemistry examination items. The findings were intended to inform the creation of new assessment items that are more explicit in their incorporation of science practices. The authors argued for building upon the implicit incorporation of science practices already present in many of the items that were analysed. While the recent use of the terminology of "scientific practices" has been prevalent through the impact of the NGSS in the international education landscape, the terminology "practical science" has been widely used in the curriculum, exam board specifications and research traditions in England for several decades. For example, the Royal Society has used the terms "practical science" as "…a shorthand for the full programme of experimental and investigative activities (including fieldwork) conducted as part of science education in schools and colleges" (House of Lords, 2006, p. 63). Other terms have been used in international context to indicate similar ideas. For example, in the USA the National Science Education Standards promoted the use of "inquiry" (NRC, 1996, p. 31) until recently when "scientific practices" became more commonplace (NGSS Lead States, 2013). The particular characterisations of scientific practices and inquiry as well as practical science may differ across national contexts and research traditions. However, there is one aspect of the science curriculum that many international science educators have raised concerns about. Authors have noted that "laboratory activities have engaged students principally in following ritualistic procedures to verify conclusions previously presented by textbooks and teachers" (Lunetta, Hofstein, & Clough, 2007, p. 396).

As Donnelly and colleagues argued, school science in England presents a limited set of experiments that do more service to students receiving higher marks than to promote understanding of science or engagement in creative inquiries (Donnelly, Buchan, Jenkins, Laws, & Welford, 1996). According to major reviews of research literature undertaken by Reiss, Abraham and Sharpe (2012) and Dillon (2008), there is strong evidence that the assessment regime in England and Wales has had a major impact on the amount and variety of practical work that many teachers carry out. There is growing concern that the amount and quality of practical work carried out in schools suffer as a result of the impact of the national tests in science.

One of the key conclusions of Dillon's report was that pupils "fail to perceive the conceptual and procedural understandings that were the teachers' intended goals for the laboratory activities. Students spend too much time following 'recipes' and, consequently, practising lower level skills" (p. 8).

This chapter examines the context of chemistry assessments in high-stakes examinations at secondary level in England. In particular, it focuses on the nature of practical work promoted in chemistry examination papers. Following a review of the national curriculum context, attention is turned to a discussion of the underlying scientific methods that are embedded in practical work. There has been widespread concern that the notion of "scientific method" in school science is based on a mythical account of how methods work in science, particularly in relation to the linear progression from a hypothesis to experimentation (e.g. McComas, 2014). A framework of scientific methods (i.e. Brandon, 1994) is presented to illustrate different categories of methods used in science including manipulative and non-manipulative approaches to hypothesis testing which challenge narrow accounts of how the scientific method works. In this context, manipulation refers to the manipulation or changing of variables, not equipment or resources in a procedural sense. The framework on scientific methods is subsequently applied as an analytical tool to investigate what methods underpin chemistry examination papers from one of the biggest and influential examination boards in England.

Practical Science in Curriculum and Assessment in England

The terminology of "practical science" has centred in curriculum initiatives (e.g. DfE, 2014) and research investigations (e.g. Abrahams, Reiss, & Sharpe, 2013) in England for numerous years. The term "practical work" is used as an overarching term that refers to any type of science teaching and learning activity in which pupils are involved in manipulating and/or observing objects and materials in order to understand how science works. Wilson (2013) considers the objectives of doing practical science to include motivation for pupils, consolidation of theory, development of manipulative skills and understanding of data handling, but also cautions:

> However, there has been some serious questioning of the efficacy of work done in science practical lessons. There is also a consensus that the type of isolated practical tasks carried out for assessment purposes at GCSE do not enable students to find out how scientific enquiry really works. As such, the forms of summative assessment currently used at GCSE are considered not to support the aims of practical science. (p. 4)

GCSE stands for "General Certificate of Secondary Education". It is an academic qualification generally taken in a number of subjects by pupils in secondary education in England, Wales and Northern Ireland. Each GCSE qualification is in a particular subject such as chemistry and stands alone (although a set could also be pursued). Studies for GCSE examinations generally take place over two or three years depending on the subject. In recent years, the education system in England underwent a set of reforms beginning in 2010[1] that resulted in a revised curriculum for many subjects to be taught from 2015 onwards. Importantly, these reforms also brought in a new assessment regime beginning with some core subjects in 2017, with the other subjects, including the sciences, following in 2018–2019. As part of these reforms, non-examination assessments such as coursework or controlled assessments were removed from all subjects other than those where a student's performance (e.g. dance or drama, or the production of physical objects, e.g. art or Design and Technology) is the most valid expression of their subject skills and competencies. This means that the vast majority of post-reform subjects are now assessed at age 16 with GCSEs using only a set of written examination papers.

In the recent landscape, science subjects including chemistry no longer include a "hands-on" assessment of practical science skills as they did pre-reform. Instead, the final examination papers are intended to have items specifically written to indirectly assess students' knowledge and understanding of practical science (Ofqual, 2015).[2] The Office of Qualifications and Examinations Regulation advocates that at GCSE level pupils should "develop their ability to evaluate claims based on science through critical analysis of the methodology, evidence and conclusions, both qualitatively and quantitatively" (Ofqual, 2015, p. 5). Ofqual is a non-ministerial government department that regulates qualifications, examinations in England and, until May 2016, vocational qualifications in Northern Ireland.

In parallel to reform in assessment of practical science, there have been some changes in the science curricula. The component of the most recent curriculum that concerns practical science is referred to as "Working Scientifically". The Department for Education (2014) has put forward four key areas of skills to be developed through this component of the curriculum: (a) development of scientific thinking, (b) experimental skills and strategies, (c) analysis and evaluation and (d) scientific vocabulary, quantities, units, symbols and nomenclature (pp. 7–8). Examination boards have adapted Working Scientifically for instructional and assessment purposes. A key aspect of one exam board, AQA (Assessment and Qualifications Alliance), for example, is that these skills are separate but "interconnected" (AQA, 2016, p. 2). Although there are instances of alignment between curriculum and assessment goals, often innovation in assessment does not mirror innovation in curriculum design (Linn, Baker, & Dunbar, 1991). Furthermore, innovation in curriculum and assessment is not always accompanied by teacher engagement to ensure effective implementation of ambitious curricular goals (Ryder, Banner, & Homer, 2014).

Despite extensive emphasis on practical work in school science and on assessment more generally, the research literature on the assessment of practical work in science education is limited in England (Abrahams et al., 2013). First, there is different terminology related to "practical science" which can be confusing in setting assessment objectives. For example, the terms used include experimental work, prac-

---

[1] A full list of all documentation referring to the reforms is available at the time of writing from https://www.gov.uk/government/collections/gcse-as-and-a-level-reforms.

[2] The available marks for these 'practical items' must be 15% or more of the total available marks for the whole paper.

tical and enquiry skills, practical and investigative activities (QCA, 2007) as well as practical experimentation, and practical scientific methods (DfE, 2014). The ambiguity in the language around practical science is compounded with an overemphasis on the experimental method even though scientists also use other methods such as non-manipulative observations and manipulative descriptions (Erduran & Dagher, 2014). When the diversity of methods is overlooked in science education, knowledge obtained from non-experimental methods, such as non-manipulative descriptive classification (e.g. classification of elements in the Periodic Table), is likely to be viewed by teachers and pupils as less privileged or less important than rigorous hypothesis testing (e.g. investigating the effect of temperature on pressure of a gas). There is little understanding of how high-stakes examinations position “practical science” in terms of the scientific methods that they involve (e.g. non-manipulative vs. manipulative methods). Articulating the nature of methods underpinning practical science items in examinations, then, can potentially be useful in informing the design of assessments as well as the related curriculum content.

Differentiating Scientific Methods

Many practical science activities involve procedures that do not make sense from the pupils’ point of view. Mindless pursuit of procedures has typically been referred to as the “cookbook problem”. Leonard (1991) provides a description of a cookbook laboratory exercise when he writes

> This student [previously described] is the victim of the overly prescriptive laboratory investigation, typical of those used in college introductory science courses. Such laboratory experiences tend to begin with the instructor explaining to the students, often in some detail, what will happen during the exercise in an attempt to make certain that the student will carry out the exercise “correctly”.
>
> The student is then left to follow a lengthy and detailed procedure in the laboratory textbook, which will occasionally call for responses such as describing what happens with the apparatus, making a drawing, or answering a specific question in the spaces provided in the manual. The entire procedure is very prescribed, that is, the student is told what to do in a step-by-step fashion for the entire exercise. (p. 84)

In a similar vein, McPherson (2018) highlights the need for updated high school chemistry practicals that reflect the recent paradigm shifts in education, moving away from traditional cookbook practicals to authentic problems that promote student thinking and problem-solving skills. Students using cookbook practicals do not form a complete picture of what happened because they focus on each step independently. Pupils in science lessons understandably question why they are doing what they are doing. They ask such profound questions because while the purpose of an investigation may be apparent to the teachers, it is not to them. For example, they don’t understand why they have to take a measurement about volume when this measurement is so far removed from building an explanation about a phenomenon such as the mole concept. Researchers have been noting for decades that pupils need to engage in practical activities that have a closely aligned explanatory framework to give them a sense of purpose in their learning (Bransford, Brown, & Cocking, 1999).

An aspect of recipe following in practical work involves hypothesis testing in experiments. Pupils are indoctrinated into a perspective of science as hypothesis testing, with associated concepts of dependent and independent variables taught without much thinking devoted to how different questions might demand different scientific methods. Research evidence suggests that science involves many other methods than just hypothesis testing. Brandon (1994) illustrates that not all experiments rely on hypothesis testing, and that not all descriptive work is non-manipulative. He represents the connections between experiments and observations in terms of a matrix (i.e. two-by-two table) in which the nature of the investigation (experiment/observation) is related to whether it involves manipulation or not, and whether it involves hypothesis testing or parameter measurement. The nature of the investigation (experiment/observation) is related to whether or not it involves manipulation, hypothesis testing or parameter measure (see Table 9.1).

The convergence of evidence from different methods can then be used to lead to a broad explanation. It is not one method or one line of experimental or observational evidence that support complex theoretical claims but several lines of evidence need to be synthesised to bring about the level of theoretical rigour that is typically associated with established scientific knowledge. Components of evidence from these different sources drive explanatory frameworks. Erduran and Dagher (2014) identified examples of methods in chemistry (see Table 9.2). They drew on the work of Scerri (2007) who describes how Mendeleev predicted the existence of the element gallium (or eka-aluminium) through a non-manipulative description coupled with quantitative reasoning about atomic weights:

Table 9.1 Types of scientific methods (reproduced from Brandon, 1994, p. 63)

	Experiment/observation	
	Manipulate	Not manipulate
Test hypothesis	Manipulative hypothesis test	Non-manipulative
Measure parameter	Manipulative description or measure	Non-manipulative description or measure

Table 9.2 Methods related to the Periodicity of Elements (from Erduran & Dagher, 2014, p. 102)

	Manipulate	Not manipulate
Test hypothesis	Manipulative hypothesis test *E.g. Crookes' study of gases*	Non-manipulative hypothesis test *E.g. De Boisbaudran's discovery of gallium*
Measure parameter	Manipulative description or measure *E.g. Rutherford's artificial transmutation of elements*	Non-manipulative description or measure *E.g. Mendeleev's prediction of gallium*

> Mendeleev could interpolate many of the properties of his predicted elements by considering the properties of the elements on each side of the missing element and hypothesizing that the properties of the middle element would be intermediate between its two neighbors. Sometimes he took the average of all flanking elements, one on each side and those above and below the predicted element. This interpolation in two directions was the method he used to calculate the atomic weights of the elements occupying gaps in his table, at least in principle. (Scerri, 2007, p. 132)

Scerri states that it was the French chemist Emile Lecoq De Boisbaudran who subsequently "worked independently by empirical means, in ignorance of Mendeleev's prediction, and proceeded to characterize the new element spectroscopically" (Scerri, 2007, p. 135). De Boisbaudran was testing the hypothesis of the existence of a new element by spectral analysis of an ore and managed to isolate gallium through this method. Examples of chemical methods include: (a) Crookes' study of gases where pressure and voltage were used as variables in spectroscopic study of elements (e.g. Scerri, 2007, p. 251) as an example of manipulative hypothesis testing, and (b) Rutherford's artificial transmutation of elements through bombardment of nuclei with protons (e.g. Scerri, 2007, p. 253) as an example of manipulative description. All together these methods, along with numerous others, contributed to the characterisation of elements. Considering the relevance of Brandon's categories for methods in chemistry, how do examinations in school chemistry cover such categories? This is the primary question that guided the analysis of examination papers from England presented in the next section.

Methodology

The section describes an empirical investigation on chemistry examination papers from England. The main research question guiding the study was "*what methods underlie the practical chemistry items in selected papers from a major examination board in England?*" Brandon's matrix (1994) was used as an analytical tool to investigate what scientific methods are present in the items included in the examination papers. The assessments used in the analysis are typical of the post-reform question style and content that students should expect as part of their end-of-course written examinations. These materials are publicly available from examination boards, and were created to help teachers and students prepare for the summer 2018 national examinations. In England, the science subjects have two ability-based tiers: Foundation and Higher. The higher tier papers have been used for this analysis as the majority of students sit the higher tier papers.

Example items from the two papers are used to illustrate how the Brandon's matrix applies to different practical chemistry cases. Consider metal displacement reactions. One question asks that the students will vary a metal in a solution and observe the difference in the reactivity. This item involves the observation of the reaction and can thus be considered a kind of descriptive observation. There is no strict sense of making measurement. In this practical, there is manipulation of what metal is used in which solution, and an observation of the outcome. These features make the question an example of manipulative description. In contrast, an examination item includes measurement of parameters but does not involve any manipulation of variables. The practical on burning of magnesium requires observations without altering any variables. However if the practical was based on burning magnesium with different amounts of oxygen, then there would have been manipulation of the amount of oxygen as a variable. An example of manipulative hypothesis testing is about rusting of iron. One item consists of an experiment that is set up to test a hypothesis about what influences rusting. The variables include dry air, oil and painting in different combinations. There are implicit hypotheses about which of these variables contribute to rusting and in which combination. The mass of the nails is measured, and eventually the practical is meant to get the students to construct the equation of the chemical reaction of rusting. An example of a non-manipulative hypothesis testing is the prediction of the characteristics of an element in the Periodic Table. One could produce a hypothesis about what the chemical and physical properties of an element would be. In this sense, there would be hypothesis testing. However, there would be no manipulation as such, simply an evaluation of what is known and prediction of what could be. Hence, this item would count as non-manipulative hypothesis testing.

At the beginning of the study, the authors individually analysed the examination papers using Brandon's Matrix as the analytical framework by making judgments such as the above description. Each item on the specimen chemistry examination papers was analysed, and once completed the authors compared their results and percentage agreement was obtained for this first round of analysis, which was about 71%. The authors discussed the disagreements, and the unit of analysis was agreed by focusing on the key criteria: manipulation of variables (or not) and presence of hypothesis testing (or not). During this consultation phase, in order to agree on the unit of analysis some negotiation between the authors as raters was needed when difficultly occurred in classifying the items. Sometimes, the overall question discussed a particular practical investigation that could be classed as a manipulative hypothesis test. However, the item that followed asked a very specific question would fall into the non-manipulative parameter measurement. Or, similarly an item may be a knowledge question, not including any reference to practical work. An example would be reference to a particular practical activity like "*zinc nitrate can be made by reacting zinc oxide with nitric acid,* HNO_3" which could be classified as a Brandon category. However, the item might be followed up with a further part such as "*write a balanced symbol equation for this reaction*" which would not fit into Brandon's categories. Although the item relates to practical science and the reaction of two reactants to form a product (non-manipulative parameter measurement), the items that follow ask the candidate to write a balanced equation which are not directly related to practical work and therefore such occurrences were not classified.

Using the above procedure for categorisation, the authors reviewed the results further bearing in mind methodological concerns about coder agreement (e.g. Boyatzis, 1998). The marking schemes were also consulted to understand what was required to answer the item, which helped with the agreed upon unit of analysis and categorisation of the items. Percentage agreement was once again obtained which was higher at 98%. The percentage agreement is obtained by examining the categories each author placed the items. When they were categorised the same, a zero was recorded. Where there was no match, the number one was recorded. All the zeros were counted and put over the number of items for each paper and multiplied by 100 to get the percentage agreement. Miles and Huberman (1994) suggest anything over 80% agreement is a good indicator of reasonable reliability.

Results and Findings

The results of the analysis of the exam papers are provided in Tables 9.3 and 9.4. In the case of Exam Paper 1, the total number of items on all chemistry topics was 47. Of these, 18 were dedicated to practical chemistry. The distribution of the items relative to Brandon categories was as follows: 1 was on manipulative hypothesis testing, 2 on manipulative parameter measurement, 1 on non-manipulative hypothesis testing and 14 on non-manipulative parameter measurement. The results indicate that for both Paper 1 and Paper 2, non-manipulative parameter measurement was the method assessed at the highest frequency. In both papers, manipulative hypothesis testing was the category with the lowest frequency of items. Furthermore, the mark allocation was the highest in both papers in the non-manipulative parameter measurement category. However, Paper 2 had a higher percentage of items and marks dedicated to manipulative parameter measurement as compared to Paper 1. The trends in the distribution of the categories illustrate a relatively uniform distribution within each paper's items and allocated marks, although the relative distribution was different across the two papers (see Fig. 9.1). Paper 2 had higher percentage of manipulative parameter measurement in both items and the marks as compared to those from Paper 1.

Overall, the pattern suggests consistency between the items allocated to each category and marks although in Paper 2, there were more marks allocated to manipulative hypothesis testing even though the relative frequency of this category was the lowest in the items. In other words, in the case of Paper 2 there was relatively more

Table 9.3 Results for Exam Paper 1

Total number of items: 47			Total marks: 100 marks	
	M/HT	M/PM	Non-M/HT	Non-M/PM
No. of items	1	2	1	14
% Items	2	4	2	30
Marks	2	4	3	24
% Marks	2	4	3	24

Table 9.4 Results for Exam Paper 2

Total number of items: 40			Total marks: 100 marks	
	M/HT	M/PM	Non-M/HT	Non-M/PM
No. of items	1	5	3	11
% Items	2.5	12.5	7.5	27.5
Marks	6	15	7	27
% Marks	6	15	7	27

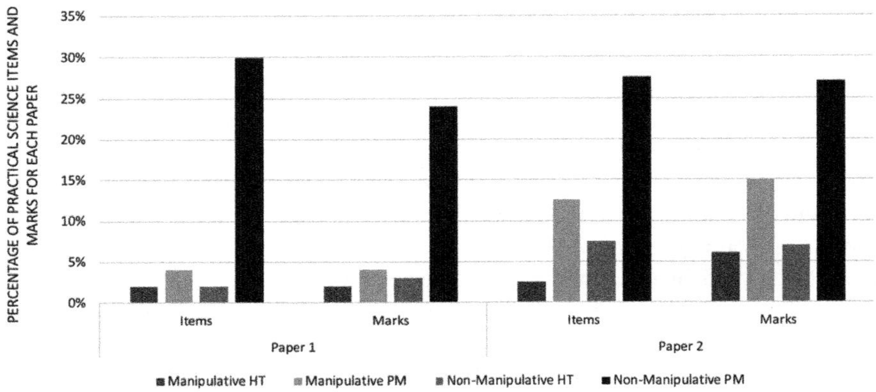

Fig. 9.1 Distribution of items and marks in Exam Papers 1 and 2

items dedicated to non-manipulative parameter measurement (i.e. 11 or 27.5%) as compared to manipulative parameter measurement (i.e. 5 or 12.5%). However, the relative distribution of marks was not consistent. The items and marks dedicated to non-manipulative parameter measurement (i.e. 27.5 and 27% respectively) as compared to manipulative parameter measurement (i.e. 12.5 and 15%) suggest that more marks were dedicated to manipulative parameter measurement as compared to the number of items covered. This observation may suggest an assumption that manipulative parameter measurement is considered to demand a higher cognitive ability thus deserving more marks (e.g. Cullinane & Liston, 2016).

Conclusions and Discussion

The results from the empirical investigation of examination papers from one English examination board indicate that some scientific methods are weighted more than others particularly in the case of the manipulative parameter measurement category. For this category, there was also a higher percentage of marks. The opposite was evident for non-manipulative parameter measurement. The questions included a higher percentage of the latter category although the percentage of marks was lower. The overall trends indicate that there is disproportionate emphasis on different methods in the examination questions of both papers. This finding raises questions about secondary students' experiences in learning of methods in chemistry. If the majority of examination questions on practical chemistry emphasise to a greater extent one kind of method, it is likely that teaching approaches will dominate this method as well. Diversifying the types of questions in examinations will imply that chemistry teachers will be more inclined to present students a more a balanced experience of chemistry through practical activities.

The use of Brandon's matrix has informed the theoretical background of the chapter. It highlighted the types of methods that can underpin practical chemistry activities. Considering some ambiguities around the terminology of "practical science" (Abrahams et al., 2013), clarifying the kinds of methods used in science may help curriculum developers, writers of assessments and chemistry teachers to have a more representative and inclusive account of chemistry practical activities. As previously stated, various terms have been used in relation to practical science such as *experimental work, practical and enquiry skills, practical and investigative activities* (QCA, 2007) and *practical experimentation,* and *practical scientific methods* (DfE, 2014). Brandon's matrix introduces a level of characterisation of methods that would encompass all of these approaches and draw attention to the fundamental reasons for why chemists do practical work. This is because Brandon's categories highlight the various ways through which scientists generate knowledge. Furthermore, it focuses attention to what is significant to unravel in terms of how different methods work in science, thus removing the traditional problems associated with the "cookbook problem" where students engage in practical activities mindlessly. The adaptation of Brandon's matrix as an analytical tool served empirical purposes because it allowed for investigating the trends in the examination items. Hence, the chapter also illustrates how a theoretical framework can be adapted for methodological purposes so as to help investigate research problems in chemistry education.

The chapter raises questions about the design of assessments for evaluating students' learning of practical chemistry. Ensuring a balanced approach to the representation of methods in the examination items will likely lead to a more diverse range of methods being used in teaching, thus improving students' understanding of how different methods work in chemistry. If some methods are promoted to a larger extent, it's questionable how students can be taught either as future chemists or chemically literate citizens to understand how chemists actually do chemistry. Future research and development efforts could focus on the design of assessments that consider further the cognitive demands placed on the students in different types of methods underpinning practical chemistry examination questions. Furthermore, the marks allocated to different examination items can be further investigated to identify how they can be more allocated for a fair indication of understanding different methods underpinning practical chemistry.

Acknowledgements The study reported in this paper was conducted in the context of Project Calibrate based at University of Oxford. The authors acknowledge the funding support from the Wellcome Trust (grant number 209659/Z/17/Z).

References

Abrahams, I., Reiss, M. J., & Sharpe, R. M. (2013). The assessment of practical work in school science, *Studies in Science Education, 49*(2), 209–251.

Achieve, Inc. (2013). *Next Generation Science Standards*. The National Academies Press: Washington, DC.

AQA. (2016). *GCSE combined science: Synergy*. Available at https://filestore.aqa.org.uk/resources/science/specifications/AQA-8465-SP-2016.PDF. Accessed 26 Feb 2019.

Boyatzis, R. E. (1998). *Transforming qualitative information: Thematic analysis and code development*. Thousand Oaks, CA: Sage.

Brandon, R. (1994). Theory and experiment in evolutionary biology. *Synthese, 99,* 59–73.

Bransford, T. D., Brown, A. L., & Cocking, R. R. (1999). *How people learn: Brain, mind, experience and school*. Washington, DC: National Academy Press.

Cullinane, A., & Liston, M. (2016). Review of the Leaving Certificate biology examination papers (1999–2008) using Bloom's Taxonomy—An investigation of the cognitive demands of the examination. *Irish Educational Studies, 35*(3), 249–267.

Department for Education. (2014). *National curriculum in England: Science programmes of study. Key Stage 4*. London.

Dillon, J. (2008). *A review of the research on practical work in school science*. London: Royal Society.

Donnelly, J., Buchan, A., Jenkins, E., Laws, P., & Welford, G. (1996). *Investigations by order: Policy, curriculum and science teachers' work under the Education Reform Act*. Nafferton: Studies in Education.

Erduran, S., & Dagher, Z. (2014). *Reconceptualising the nature of science for science education: Scientific knowledge, practices and other family categories*. Dordrecht: Springer.

House of Lords Science and Technology Committee. (2006). *Tenth report of session 2005–06 science teaching in schools*.

Leonard, W. H. (1991). A recipe for uncookbooking laboratory investigations. *Journal of College Science Teaching, 21*(2), 84–87.

Linn, R. L., Baker, E. L., & Dunbar, S. B. (1991). Complex performance-based assessment: Expectations and validation criteria. *Educational Researcher, 20*(8), 15–21.

Lunetta, V. N., Hofstein, A., & Clough, M. P. (2007). Teaching and learning in the school science laboratory. An analysis of research, theory, and practice. In S. K. Abell & N. G. Lederman (Eds.), *Handbook of research on science education* (pp. 393–431). Mahwah, NJ: Lawrence Erlbaum Associates.

McComas, W. F. (2014). Scientific method (Scientific methodology). In W. F. McComas (Ed.), *The language of science education*. Rotterdam: Sense Publishers.

McPherson, H. (2018). Transition from cookbook to problem-based learning in a high school chemistry gas law investigation. *Teaching Science, 64*(1), 47–51.

Miles, M. B., & Huberman, M. (1994). *Qualitative data analysis: A sourcebook of new methods* (2nd ed.). Beverly Hills, CA: Sage Publications.

National Research Council (NRC). (1996). *National science education standards*. Washington, DC: National Academy Press.

Ofqual. (2015, July). GCSE subject level conditions and requirements for single science (Biology, Chemistry, Physics). https://assets.publishing.service.gov.uk/government/uploads/

system/uploads/attachment_data/file/600867/gcse-subject-level-conditions-and-requirements-for-single-science.pdf. Accessed December 2018.

Qualifications and Curriculum Authority (QCA). (2007). *Science: Programme of study for key stage 3 and attainment targets*. Archived version https://webarchive.nationalarchives.gov.uk/20080610180700/, http://curriculum.qca.org.uk/key-stages-3-and-4/subjects/science/keystage3/index.aspx.

Reed, J. J., Brandriet, A. R., & Holme, T. A. (2017). Analyzing the role of science practices in ACS exam items. *Journal of Chemical Education, 94*(1), 3–10.

Reiss, M., Abrahams, I., & Sharpe, R. (2012). *Improving the assessment of practical work in school science*. London: Gatsby Foundation.

Ryder, J., Banner, I., & Homer, M. (2014). Teachers' experiences of science curriculum reform. *School Science Review, 95*(352), 126–130.

Scerri, E. R. (2007). *The periodic table: Its story and its significance*. New York: Oxford University Press.

Wilson, A. (2013). *The assessment of practical science: A literature review*. Cambridge: Assessment Research and Development.

Chapter 10
Deciphering Students' Thinking on Ionisation Energy: Utilising a Web-Based Diagnostic Instrument

Kim Chwee Daniel Tan, Keith S. Taber, Yong Qiang Liew and Kay Liang Alan Teo

Abstract Chemistry is a difficult subject to learn, in part because explanations for chemical phenomena are abstract, involving the interactions of invisible particles. Thus, students may construct ideas and explanations of chemical phenomena that are inconsistent with those accepted by the scientific community; these ideas may be resistant to change as they are based on the ways students perceive how and why things behave. It is important to identify students' alternative conceptions so that measures can be taken to help students construct more scientifically acceptable concepts. Studies have been conducted to determine students' alternative conceptions over a wide range of topics in chemistry, including ionisation energy. In a previous study, a pen-and-paper two-tier multiple choice diagnostic instrument on ionisation energy was developed for use in the classroom. This chapter describes the further development and use of a web-based version of the instrument to determine students' thinking in the topic. Such web-based instruments are easily made available on the internet, readily administered to students and the results can be automatically collated, saving the time and effort of researchers as well as teachers. This is one way in which the use of educational research by teachers can be facilitated, impacting practice in the classroom.

K. C. D. Tan (✉)
National Institute of Education, Nanyang Technological University, Singapore, Singapore
e-mail: daniel.tan@nie.edu.sg

K. S. Taber
Faculty of Education, University of Cambridge, Cambridge, UK
e-mail: kst24@cam.ac.uk

Y. Q. Liew
Millennia Institute, Singapore, Singapore
e-mail: liew_yong_qiang@moe.edu.sg

K. L. A. Teo
Hwa Chong Institution, Singapore, Singapore
e-mail: alanteo@hci.edu.sg

M. Schultz et al. (eds.), *Research and Practice in Chemistry Education*,
https://doi.org/10.1007/978-981-13-6998-8_10

Introduction

Learners actively construct their own knowledge rather than passively receive knowledge, and what they already know has a major impact on construction of new knowledge (Ausubel, 1968; Driver & Easley, 1978; Driver & Oldham, 1986; Taber, 2013; Treagust, Duit, & Fraser, 1996; Ye & Lewis, 2014). Their existing knowledge arises from what they experience in their everyday life through their senses, use of language, interactions with others, cultural background as well as through mass media and formal instruction (Adadan & Savasci, 2012; Duit & Treagust, 1995; Ryder & Leach, 2000). Research has shown that the knowledge and understanding that students bring with them to science lessons may be in conflict with those accepted by the scientific community (Osborne, Bell, & Gilbert, 1983). These alternative conceptions (Abimbola, 1988; Hewson, 1981) may be strongly held and resistant to change (Driver & Easley, 1978; Erman, 2017), so students may not change their original ideas and understandings even after being taught the scientific concepts (Fetherstonhaugh & Treagust, 1992). This can be because they are satisfied with their own conceptions, and so, do not see the need to change them (Duit & Treagust, 1995). They can also find the concepts taught incomprehensible because they cannot sensibly relate these concepts to their prior conceptions; the science concepts can be too complex, alien to daily life experiences or beyond their developmental level (Erman, 2017; Gabel, 1989; Solomon, 1993; Taber, 2000, 2006). In addition, the students can (mis)construct meanings from the new material such that the (mis)constructed meanings are more in line with their existing conceptions than the intended meanings (Johnstone, 2000; Osborne et al., 1983; Taber, 2000). Students' alternative conceptions seem to vary along a range of dimensions, from unitary to manifold, isolated to embedded in complex networks, explicit to tacit (Taber, 2014), inert, theory-like and consistent (Driver & Erickson, 1983; Gilbert & Watts, 1983) to labile and context-bound in response to demands at particular points in time (Claxton, 1993; Solomon, 1993).

Studies have indicated that there are many factors and interactions that influence the responses given by the students to questions asked by researchers (Claxton, 1993; Taber & Tan, 2007), and that students often provide inconsistent responses to questions involving very similar concepts (e.g. Adadan & Savasci, 2012; Palmer, 1999; Taber & Tan, 2007; Tan, Goh, Chia, & Treagust, 2002; Voska & Heikkinen, 2000). This may be due to students having multiple conceptions for a given conception (Taber, 2013) and "different conceptions are brought into play in response to different problem contexts" (Palmer, 1999, p. 639). Another proposal is that students have different interpretations due to implicit knowledge elements deriving from perceived patterns in perceptual data, or p-prims (diSessa, 1993), that they apply in an inconsistent manner; they might get an item correct in a test but then their responses to other items testing similar concepts indicate that they have alternative conceptions (Taber, 2006; Taber & Tan, 2007). This is especially so for topics which are abstract, complex and where "several potential relevant factors have to be considered and weighted" (Taber & Tan, 2007, p. 379).

Multiple Choice Instruments

Multiple choice tests are popular for diagnosing students' understanding of concepts as they enable a large number of students to be sampled in a given amount of time. These tests are easy to administer and score, and the results obtained are also easier to process and analyse compared to interview data, concept maps and responses to open-ended questions (Peterson, Treagust, & Garnett, 1989; Taber, 1999; Tan & Treagust, 1999). In a previous study on students' understanding of ionisation energy (Taber & Tan, 2007; Tan, Taber, Goh, & Chia, 2005), a two-tier multiple choice diagnostic instrument (Treagust, 1995), the Ionisation Energy Diagnostic Instrument (IEDI), was used to determine students' understanding of the concepts involved in ionisation energy. Students had to choose an answer from the first-tier options and one reason for their answer from the second-tier options. The distractor options in the second tier were derived from student alternative conceptions highlighted in preliminary studies as well as related studies in the literature. For an item with three options in the first tier and four options in the second tier, there seems to be 12 possible answer–reason combinations. However, in reality, there may be a reduced set of reasons which appear viable for a specific (first tier) answer (Griffard & Wandersee, 2001; Loh, Subramaniam, & Tan, 2014; Tan et al., 2002), decreasing the number of plausible answer–reason combinations. For example, if reasons 1 and 2 in an item fit answer A only, reason 3 fits answer B, and reasons 3 and 4 fit answer C, the number of logical answer–reason combinations is only five! A related issue is that students may treat their choices in the two tiers as if they are independent of each other, resulting in answer–reason combinations (e.g. choosing reason 3 or 4 for answer A) which are 'illogical' (at least to the researchers) and difficult to analyse.

Guessing, when students do not know the answer, makes it difficult for researchers to differentiate between actual understanding of concepts and student guessing the right answer, or actual alternative conceptions and lack of understanding/non-conceptions (Caleon & Subramaniam, 2010a; Hasan, Bagayoko, & Kelly, 1999). This can be addressed somewhat by using Certainty of Response Index (CRI) in diagnostic tests (Hasan et al., 1999). Using this protocol, respondents indicate how confident they are, on a six-point scale (0–5), of their knowledge required to answer a given question. If they indicate that they are uncertain (CRI of 0–2), it will be assumed that the answer is mainly due to guesswork, and if they indicate that they are more certain (CRI of 3–5) and the answer is wrong, it will be assumed that the student has a possible alternative conception. Confidence testing has been used by researchers to produce three-tier tests (Caleon & Subramaniam, 2010a; Brandriet & Bretz, 2014) where the respondents indicate their confidence in their answer–reason combination and four-tier tests (Caleon & Subramaniam, 2010b; Sreenivasulu & Subramaniam, 2013; Lin, 2016; Yan & Subramaniam, 2018) where they indicate their confidence in their answer and the associated reason separately.

The wIEDI

The web-based Ionisation Energy Diagnostic Instrument (wIEDI) (https://tinyurl.com/wIEDI-ICCE) [password: icce] was developed based on the IEDI using the Google Forms platform (Tan, Taber, Liew, & Teo, in press). Teachers and students can readily access instruments developed using Google Forms from the internet, and data can be downloaded as Microsoft Excel documents for further analysis. This ease of access and use can facilitate the dissemination of such research-informed material to, and uptake of such instruments by, teachers for use in their lessons. Item 1 is given in Fig. 10.1 as an example of the items in the wIEDI.

The stem and the first-tier options of all 10 items in the wIEDI are the same as those in the pen-and-paper version. To address students treating the two tiers as

For Questions 1 to 5, please refer to the statement below.
Sodium atoms are ionised to form sodium ions as follows:
$Na(g) \rightarrow Na^+(g) + e$

Once the outermost electron is removed from the sodium atom forming the sodium ion (Na^+), the sodium ion will not combine with an electron to reform the sodium atom.
(A) True.
(B) False.
(C) I do not know the answer.

Your Response: (A) True.
Reason(s) (You may choose as many options as you think are correct):
(1) Sodium is strongly electropositive, so it only loses electrons.
(2) The $Na^+(g)$ ion has a stable/noble gas configuration, so it will not gain an electron to lose its stability.
(3) The process of Na(g) losing an electron is irreversible.
(4) A very large amount of energy is required to add an electron to the $Na^+(g)$ ion.
Other: ________________

How confident are you that your answer and reason(s) are correct?
( ) Confident
( ) Not confident

Your Response: (B) False.
Reason(s) (You may choose as many options as you think are correct):
(1)The positively-charged $Na^+(g)$ ion can attract a negatively-charged electron.
(2) $Na^+(g)$ can react with an anion to reform the Na atom.
(3) $Na^+(g)$ can undergo a displacement reaction to form Na.
Other: ________________

How confident are you that your answer and reason(s) are correct?
( ) Confident
( ) Not confident

Your Response: (C) I do not know the answer.
Can you elaborate on why you do not know the answer or are uncertain of the answer?
____

(Answer: B1)

Fig. 10.1 Item 1 in the wIEDI

independent of each other and to reduce 'illogical' answer–reason combinations, the wIEDI made use of the branching function in Google Forms to present different subsets of second-tier reasons depending on the answers selected in the first tier. Thus, the second-tier options had to be reorganised into subsets and additional options had to be included so that there were at least two options in any set of second-tier reasons. For example, in the original IEDI, the same three reasons in the second tier of item 1 are presented for both the True and False answers in the first tier. In contrast, the same item in the wIEDI has seven reasons organised into two different subsets depending on the choice of answer in the first tier. Each subset contains one or more reasons originally present in the second tier of the item in the IEDI and possibly additional reasons from other similar items in the IEDI, as well as responses given in the open-ended versions of the instrument and interviews from the previous study. The students were encouraged to choose option C (I do not know the answer) if they did not know the answer to any item; this was also the case in the original IEDI. The students were then prompted to explain why they did not know the answer to allow the researchers to understand what their difficulties were. For diagnostic purposes, this was preferred to allowing students to choose one or more reasons in the second-tier options to support an answer that they did not know, or were not confident of.

In addition, each subset was configured as a checklist to allow students to choose more than one answer. If the student could not find any suitable reason or wanted to include a reason or reason(s) not offered in the second tier, they could choose the 'Other' option and type in their reason(s) in the 'Other' text box. Allowing students to choose more than one reason could address the issue of inconsistent responses to questions involving very similar concepts as they now have the opportunity to display all relevant conceptions that they have for a given situation (Palmer, 1999) or multiple p-prims utilised to answer the questions (diSessa, 1993). It was decided that there would only be two points, 'Confident' or 'Not confident' on the confidence scale to simplify students' choices, removing the need to consider the extent of confidence in their choices. The confidence scale would cover both tiers together (answer–reason combination) rather than having scales for both tiers separately as the reason options offered to the students were dependent on the answers chosen by them. Only answer–reason combinations that students indicate confidence in were used to determine the percentage of students choosing such combinations; this was to minimise the impact of guessing on the results.

The study was approved by the Nanyang Technological University Institutional Review Board and permission was given by the Ministry of Education, Singapore, to collect data in schools. The web-based diagnostic instrument underwent two trials and was revised based on the results as well as interviews with students to probe the reasoning behind their choices and their views of the instrument. The finalised version of the wIEDI was validated by seven high school teachers and two tertiary chemistry educators in terms of its accuracy of content and relevance to the high school chemistry syllabus. Eight high schools in Singapore were involved in the study from May to July 2016. Parental consent and assent from students were obtained online, and a total 274 students (257 Grade 11 and 17 Grade 12) agreed to participate in the study. The results were downloaded as Microsoft Excel documents and analysed using the Microsoft Excel program.

Students' Thinking on Ionisation Energy

Given that learners' thinking can be idiosyncratic, discussion of response combinations will be limited to those which students indicated their confidence in and reaching 5% in the survey. The previous study highlighted choices made by 10% or more students (they were not required to indicate their confidence in their responses). The students' difficulties determined in this study are similar to those in previous studies (see Table 10.1); these included the octet rule framework, stable fully-filled subshell and half-filled subshell conceptions, conservation of force thinking and relation-based reasoning (Driver, Leach, Millar, & Scott, 1996; Taber, 1999, 2003; Taber & Tan, 2007; Tan et al., 2005). Students who resorted to the octet rule framework, stable fully-filled or half-filled subshell conception seemed to believe that atoms or ions 'strive to obtain' certain electron configurations to achieve stability, and were 'reluctant' to lose these electronic configurations. The conservation of force thinking led students to consider that the nuclear attraction was shared among the oppositely charged electrons such that there was redistribution of the force associated with the nuclear charge among the remaining electrons if one or more electrons left the atom or ion. Relation-based reasoning caused students to consider only one or two factors instead of all the factors involved in the determination or comparison of ionisation energies.

Answer–reason combinations are referred to by a letter which denotes the first-tier answer and a number which denotes the associated second-tier reason. The alternative conception which is indicated by the answer–reason combination is usually presented as well (refer to Table 10.1 and the wIEDI). For example, in item 1, A2 (oct) shows the choice of answer A and reason 2, indicating the use of the octet rule framework in answering the item. There are more alternative conceptions indicated using the wIEDI than in the previous study, as exemplified by items 1–3 in Table 10.1. This was to be expected as new reason responses had to be identified to ensure that there were sufficient reasons in the second tier to cater for each first-tier answer, and students were allowed to choose more than one reason for each answer in the wIEDI. For example, in the pen-and-paper IEDI, there is the correct reason and two distractors in the second tier of item 1 catering to both 'True' or 'False' answers in the first tier. In item 1 of the wIEDI, there are four reasons in the second tier for incorrect answer, (A) True, and the correct reason as well as two other distractors in the second tier for correct reason, (B) False. Thus, the results using the wIEDI showed that 5% or more students chose options in item 1 indicating three additional alternative conceptions: (1) ionisation is endothermic (endo); (2) the sodium ion can react with other species such as anions to reform the sodium atom (rxn); (3) the sodium ion can undergo reduction reactions (which is highly unlikely) to reform the sodium atom (red). The following sub-sections describe the thinking possibly underlying students' selection of answer–reason(s) combinations.

Table 10.1 Response combinations in items 1–3 of the wIEDI indicating alternative conceptions chosen confidently by 5% or more of Grade 11 and 12 students ($n = 274$) in the current study compared with those chosen by Grade 11 and 12 students ($n' = 979$) in the previous study

Item	Reason	Alternative conception indicated by response combination	Percentage of students ($n = 274$)	Percentage of students in previous study ($n' = 979$)
1	A1 (pos)	Strongly electropositive sodium only loses electrons	8	5*
	A2 (oct)	The sodium ion has a stable/noble gas configuration, so it will not gain an electron to lose its stability	17	44
	A4 (endo)	Ionisation is endothermic	7	N.A.**
	B2 (rxn)	The sodium ion can react with an anion to reform the sodium atom	9	N.A.**
	B3 (red)	The sodium ion can undergo a displacement reaction to form the sodium atom	7	N.A.**
2	A1 (cof)	The attraction of the nucleus for the 'lost' electron will be redistributed as the • number of protons in the nucleus is the same but there is one less electron to attract, so the remaining 10 electrons will experience greater attraction by the nucleus	37	50
	A2 (cof)	• $Na^{+}(g)$ ion has one less shell compared to the Na(g) atom	14	N.A.**
	A3 (cof)	• total force of attraction by the nucleus is always divided equally between the total number of electrons	9	N.A.**

(continued)

Table 10.1 (continued)

Item	Reason	Alternative conception indicated by response combination	Percentage of students (n = 274)	Percentage of students in previous study (n' = 979)
3	A2 (red)	The sodium ion has a vacant shell which can be filled by electrons from other atoms or molecules to form the sodium atom	6	5*
	A3 (rxn)	The sodium ion has a positive charge and will react readily with other atoms or molecules to form the sodium atom	8	N.A.**
	B1 (cof)	The average force of attraction on each electron of the sodium ion is greater than that of the sodium atom as the attractive force is shared among fewer electrons	6	1*
	B2 (oct)	The outermost shell of the sodium ion has achieved a stable octet/noble gas configuration	36	64
	B3 (ffs)	All subshells are filled fully with electrons for the sodium ion whereas the sodium atom has one subshell with a lone electron that give it its instability	26	N.A.**

Notes *Indicates that the percentage of students is less than 10% in the previous study (Tan et al., 2005)
N.A.**Indicates that the option was not present in the previous study

Responding to Context of the Question

The 'octet rule framework' reason was chosen by 5% or more of 274 students in five items (see Table 10.2). The figures represent the number of students who confidently chose a particular option in all possible combinations. For example, 7% of 274 students confidently chose the 'octet rule framework' reason (B1) in item 5. This included students who chose option B1 only, as well as the combinations (B1, B2) only, (B1, B3) only, (B1, B2, B3) only and (B1, B2, Other) only. The students' choices were not consistent as shown by the different percentages for the five items even though they could choose the 'octet rule framework' reason in addition to the other reasons that they believed supported their answer in the first tier.

Cross-tabulation was also used to study the consistency of the students' choice (Tan et al., 2002) across items in the wIEDI. To illustrate the computation involved, the percentage (8%) for the cross-tabulation of Q1 (A2) and Q3 (B2) (see Table 10.2) was derived from the number of students who chose one of these combinations [A2 only, (A1, A2) only, (A1, A2, A3) only, (A1, A2, A4) only, (A2, A3) only, or (A2, A4) only] in item 2 and one of these combinations [B2 only, (B1, B2) only, (B1, B2, B3) only, or (B2, B3) only] in item 3. There were three instances of consistent choices of the 'octet rule framework' reasons over two items by 5% or more of the respondents, but none involved items 5 and 7. Palmer (1999) suggests that students select answers in response to particular situations based on an "if…then" understanding of the content, and so, may not realise that there are inconsistencies in their responses (Palmer, 1999). There was explicit focus on the Na^+ ion in items 1–4, so its electronic configuration could have featured strongly in the students' considerations, whereas items 5 and 7 compared the first ionisation energies of sodium and magnesium, and sodium and aluminium, respectively. Even then, the cross-tabulation percentages (5–10%) across items 1, 3 and 4 and the percentages for the 'octet rule framework' reasons in the items individually (17–36%) indicated that the students did not use the reason all the time in similar situations to justify their answers though they could choose the reason, as well as others, for their answers.

Similarly, the 'stable fully-filled subshell conception' reasons in items 3, 4, 5 and 6 were confidently chosen by more than 5% of 274 students, but the figures for items 5 (11%) and 6 (5%) were less than half of those in items 3 (26%) and 4 (25%) (see Table 10.3). Cross-tabulation of the 'stable fully-filled subshell conception' reasons over the four items showed only one instance of consistency and it involved items 3 and 4 (9%). Again, it seemed that thinking involving electronic configuration was more strongly featured in items 3 and 4 where the Na^+ ion was foregrounded; the Na^+ ion has a fully-filled outermost $2p$ subshell. Items 5 and 7 involved the comparison of magnesium (which has a fully-filled outermost $3s$ subshell) with sodium and

Table 10.2 'Octet rule framework' options confidently chosen by 5% or more students

Option	Percentage of students ($n = 274$)
Q1 (A2)	17
Q3 (B2)	36
Q4 (A1)	22
Q5 (B1)	7
Q7 (B1)	5
Q1 (A2), Q3 (B2)	8
Q1 (A2), Q4 (A1)	5
Q3 (B2), Q4 (A1)	10

Table 10.3 'Stable fully-filled subshell conception' options confidently chosen by 5% or more students

Option	Percentage of students ($n = 274$)
Q3 (B3)	26
Q4 (A4)	25
Q5 (B2)	11
Q6 (A1)	5
Q3 (B3), Q4 (A4)	9

aluminium, respectively, but the case for magnesium did not resonate as much with the students as the case for the Na^+ ion. Perhaps the electronic configuration of the Na^+ ion and its 'stability' was more familiar to the students due to the use of sodium chloride as an example in teaching of chemical (ionic) bonding from Grades 8 onwards, whereas subshells and orbitals were only introduced in Grade 11.

Analogous Reasoning

However, cross-tabulation of students' confident choices of the 'octet rule framework' and 'stable fully-filled subshell conception' reasons within and across items 1, 3 and 4 indicated several instances of confident and consistent choices of both reasons (see Table 10.4). For example, in item 3, 25% chose B2 (oct) and B3 (ffs) in all combinations, compared to 36% who chose B2 (oct) and 26% who chose B3 (ffs); this meant that almost all the students who chose B3 (ffs) also chose B2 (oct). A lower percentage of students (12%) confidently chose A1 (oct) and A4 (ffs) in item 4 in all combinations compared to the 22% who chose A1 (oct) and 25% who chose A4 (ffs). Across items 3 and 4, 5% choose both 'octet rule framework' reasons [Q3 (B2), Q4 (A1)] as well as both 'stable fully-filled subshell conception' reasons [Q3 (B3), Q4 (A4)]! This result was not surprising given that the Na^+ ion had an octet of electrons in the outermost shell and a fully-filled 2p subshell. In the previous study, Tan et al. (2005) argued that the octet rule framework led readily to the stable fully-filled subshell conception as they were analogous to each other (Duit, 1991) in terms of 'stable' electronic structures. Thus, both sets of reasons were attractive to students, especially in items 3 and 4.

Three 'stable half-filled subshell conception' options were confidently chosen by 5% or more of the students in items 8 and 9 (see Table 10.5), but cross-tabulation did not show any instance of consistent choice of the reason by 5% or more students. In addition, there were no instances of 5% or more students choosing reasons indicating the stable half-filled subshell conception and those indicating the octet rule framework or the stable fully-filled subshell conception. Although it was somewhat analogous to the octet rule framework and the stable fully-filled subshell conception in terms of symmetry , the stable half-filled subshell conception options were not

Table 10.4 Examples of cross-tabulations of the 'octet rule framework' and 'stable fully-filled subshell conception' options in items 1, 3 and 4

Octet rule framework	Stable fully-filled subshell conception	Percentage of students ($n = 274$)
Q3 (B2)	Q3 (B3)	25
Q4 (A1)	Q4 (A4)	12
Q1 (A2)	Q3 (B3)	5
Q3 (B2)	Q4 (A4)	12
Q1 (A2), Q3 (B2)	Q3 (B3)	5
Q3 (B2), Q4 (A1)	Q3 (B3)	8
Q3 (B2)	Q3 (B3), Q4 (A4)	9
Q3 (B2), Q4 (A1)	Q3 (B3), Q4 (A4)	5

Table 10.5 'Stable half-filled subshell conception' options confidently chosen by 5% or more students

Option	Percentage of students ($n = 274$)
Q8 (B1)	8
Q9 (A2)	7
Q9 (A4)	8

very attractive to students. Perhaps they were lacking in terms of 'completeness' in the filling of the subshell, thereby making them less 'analogous', attractive and viable to students.

Multiple Conceptions

Allowing students to choose more than one reason for an answer in the wIEDI provided direct evidence of students' possible multiple conceptions (Taber, 2013) or activation of different *p*-prims, this time in response to the context of the same question. Table 10.5 shows many instances of students choosing two or more reasons in item 4 of the wIEDI indicating different alternative conceptions (octet rule framework, conservation of force thinking and stable fully-filled subshell), and even the correct reason as well as reasons indicating alternative conceptions. The sum of the percentages of students in the last column of Table 10.6 exceeded 100% because the figures were computed for the options chosen in all possible combinations. For example, the percentage of students (9%) confidently choosing Q4(A1) and Q4(A2) was computed from the number of students choosing [(A1, A2) only, (A1, A2, A3*) only, (A1, A2, A3, A4) only, and (A1, A2, A3, Other) only], while the percentage of students (5%) confidently choosing Q4(A1), Q4(A2), Q4(A3*) and Q4(A4) was computed from the number of students choosing (A1, A2, A3*, A4) only. Thus both computations included the students who chose the combination (A1, A2, A3*, A4) only.

The correct reason, A3*, was included in the confident choices made by 58% of the students, but only 13% chose the correct reason A3 by itself. This suggests that choosing the correct answer does not mean that a student has no alternative conception. It also shows the disadvantage of allowing students to decide on only one choice in instruments which are designed to diagnose students' difficulties and alternative conceptions. In item 4, students had to explain why the second ionisation energy of sodium was greater than its first ionisation energy. The correct answer and the three other distractors did not contradict each other in this situation, and each could be applicable depending on the students' (mis)understanding of the concepts involved. Thus, various combinations of reasons were attractive to students, with 5% even confidently choosing all four options!

Relation-Based Thinking

Two sets of 'relation-based thinking' (Driver et al., 1996) responses which 5% or more students confidently chose are given in Table 10.7. The 'electron is further away' reason attracted 16% of the students who confidently chose A2 (efa) in item 6 which stated that the first ionisation energy of magnesium is greater than that of aluminium because the 3*p* electron of aluminium is further from the nucleus compared to the 3*s* electrons of magnesium. This figure was more than the 7% who confidently chose the correct reason A3*. The differences in nuclear charge of magnesium and aluminium, as well as the screening effect/repulsion of other electrons present in the

Table 10.6 Examples of cross-tabulations of the 'octet rule framework' and 'stable fully-filled subshell conception', 'conservation of force thinking' and correct answer choices in items 4

Octet rule framework	Conservation of force thinking	Correct answer	Stable fully-filled subshell conception	Percentage of students (n = 274)
Q4 (A1)	Q4 (A2)			9
Q4 (A1)		Q4 (A3*)		18
	Q4 (A2)	Q4 (A3*)		33
	Q4 (A2)		Q4 (A4)	12
		Q4 (A3*)	Q4 (A4)	21
Q4 (A1)	Q4 (A2)	Q4 (A3*)		9
Q4 (A1)		Q4 (A3*)	Q4 (A4)	10
	Q4 (A2)	Q4 (A3*)	Q4 (A4)	11
Q4 (A1)	Q4 (A2)	Q4 (A3*)	Q4 (A4)	5

respective atoms were not considered. In addition, Tan et al. (2005) suggested that students might have remembered "a comparison presented in the context of a single atomic system (e.g. 3*s* of sodium compared with 3*p* of sodium), and expecting the same pattern when the comparison is made between different systems (e.g. 3*s* of magnesium compared with 3*p* of aluminium)" (p. 189). However, in this study, five of the seven teachers who validated the wIEDI expressed their concerns that their students were taught that the "3*p* electron of aluminium is further from the nucleus compared to the 3*s* electrons of magnesium" was an acceptable explanation for the higher first ionisation energy of magnesium compared to that of aluminium. Thus, students might choose reasons which were aligned with the curriculum explanations rather than the expected reason, so they could not be faulted for their choices.

A similar reason, A3 (efa), in item 7, which stated that the first ionisation energy of sodium was greater than that of aluminium because the 3*p* electron of aluminium was further away from the nucleus was confidently chosen by 12% of the students. However, this was a smaller figure compared to the 29% who confidently chose the correct reason, B2*, possibly because A3 (efa) was a justification for an incorrect answer and more students confidently chose the correct answer B (63%) than the incorrect answer A (32%). The 'electron further away' reasons in both items 6 and 7 were consistently and confidently chosen by 7% of the students.

There were three instances where 5% or more of the students confidently chose the 'ionisation energy always increases across a period' reasons (see Table 10.8). Cross-tabulations showed four instances where students confidently chose the 'ionisation energy always increases across a period' reason across items 7, 8 and 10. For example, 8% of the students confidently chose B3 (IEp) in item 7 and B3 (IEp) in item 8, and 5% confidently chose the reason across items 7, 8 and 10. The 'ionisation energy always increases across a period' reason was developed to provide sufficient options to several items in the present study, so it did not appear in the previous study. This underlying basis for students choosing this reason could be the thinking that in the same period, shielding by the inner shell electrons was similar but nuclear charge of the elements increases across a period—the effect of repulsion between electrons in the same subshell was not considered.

Table 10.7 'Electron is further away' options confidently chosen by 5% or more students

Option	Percentage of students ($n = 274$)
Q6 (A2)	16
Q7 (A3)	12
Q6 (A2), Q7 (A3)	7

Table 10.8 'Ionisation energy always increases across a period' options confidently chosen by 5% or more students

Option	Percentage of students (n = 274)
Q7 (B3)	13
Q8 (B3)	9
Q10 (B2)	13
Q7 (B3), Q8 (B3)	8
Q7 (B3), Q10 (B2)	7
Q8 (B3), Q10 (B2)	6
Q7 (B3), Q8 (B3), Q10 (B2)	5

Conclusions and Implications for Practice

The web-based diagnostic instrument facilitates branching from the first-tier answers to their associated second-tier reasons, as well as allows students to choose more than one reason for their answers. It also allows students to explain why they do not know the answer to a question and to write their own reasons if they believe that the reasons were not represented in the second-tier options. It is possible, but rather challenging, to design an equivalent pen-and-paper instrument that can incorporate these functions. Another big advantage the web-based instrument has over the pen-and-paper version is that the students' answers can be readily collated and downloaded in the Microsoft Excel format and be further processed for analysis. Optical mark readers used to scan students' answers in the pen-and-paper version may not be able to accept/process students' selection of more than one reason for an answer, and what students write may have to be manually entered into a software for collation and analysis.

As only reasons in the second tier which were confidently chosen by students were considered, guessing by students should affect the results of the present study to a lesser extent compared to the results of the previous study in which students were not required to state their confidence in their choice of reasons. Allowing students to choose more than one reason for an answer also showed how they responded to the contexts of the various questions and used analogous and relation-based thinking, as well as provided evidence of students' possible multiple conceptions. The results showed that there were many instances in which 5% or more chose two or more reasons indicating different alternative conceptions, and even the correct reasons as well as one or more reasons indicating alternative conceptions. Thus, it has to be reiterated that it does not mean that students who are able to choose the correct answer do not have any alternative conception, and that there are disadvantages of restricting students to only one answer or answer-reason option on items in diagnostic instruments. Cross-tabulation of students' choices across items highlighted the (in)consistency of students' choices of reasons and the contexts in which the different conceptions came into play.

There is the important issue of curriculum explanations taught to students. In Singapore, the Coulomb's Law may not be explicitly used in A-level chemistry to explain the attractive force between the nucleus and the electrons in an atom or ion; students studying A-level physics learn the law in electrostatics but it may not have been referred to in the context of ionisation energy by chemistry teachers. The omission can cause problems as students may use the conservation of force thinking readily as it has "an intuitive attraction to many students" (Taber, 2003, p. 156), or resort to the octet rule framework, extending it to include the stable fully-filled and half-filled subshell thinking, to answer ionisation energy questions. Thus, teachers need to be aware of "how students apply invalid explanatory principles based on the inherent stability of octets or full shells, and on notions of sharing-out of nuclear attraction" (Tan et al., 2005, p. 193) and challenge any related explanations to help students appreciate the limitations of such explanations. They should also help the students to reformulate their answers using the appropriate concepts and technical language. The use of relation-based thinking seems to suggest that students have difficulty coordinating multiple factors in the comparison of ionisation energies of different atoms. Tan et al. (2005) suggested teachers should model the thinking process, "identifying the potential factors at work, and so making decisions about which need to be taken into account in any particular case" (p. 194) to help students realise that they have to consider the effects of all relevant factors rather than focusing on only one or two factors.

The ease of the administration of the wIEDI and the processing of data, as well as the information that it can provide in the formative assessment of students' understanding of ionisation energy suggests that it can make a valuable contribution in the classroom, enabling research to impact practice.

Acknowledgements The authors would like to acknowledge the contribution of the teachers, students and school leaders involved in facilitating this study, enabling the research to be successfully completed.

References

Abimbola, I. O. (1988). The problem of terminology in the study of student conceptions in science. *Science Education, 72*(2), 175–184. https://doi.org/10.1002/sce.3730720206.

Adadan, E., & Savasci, F. (2012). An analysis of 16–17-year-old students' understanding of solution chemistry concepts using a two-tier diagnostic instrument. *International Journal of Science Education, 34*(4), 513–544. https://doi.org/10.1080/09500693.2011.636084.

Ausubel, D. (1968). *Educational psychology: A cognitive view*. New York: Holt, Rinehart & Winston.

Brandriet, A. R., & Bretz, S. L. (2014). Measuring meta-ignorance through the lens of confidence: Examining students' redox misconceptions about oxidation numbers, charge, and electron transfer. *Chemistry Education Research and Practice, 15*(4), 729–746. https://doi.org/10.1039/C4RP00129J.

Caleon, I., & Subramaniam, R. (2010a). Development and application of a three-tier diagnostic test to assess secondary students' understanding of waves. *International Journal of Science Education, 32*(7), 939–961. https://doi.org/10.1080/09500690902890130.

Caleon, I. S., & Subramaniam, R. (2010b). Do students know what they know and what they don't know? Using a four-tier diagnostic test to assess the nature of students' alternative conceptions. *Research in Science Education, 40*(3), 313–337. https://doi.org/10.1007/s11165-009-9122-4.

Claxton, G. (1993). Minitheories: A preliminary model for learning science. In P. J. Black & A. M. Lucas (Eds.), *Children's informal ideas in science* (pp. 45–61). London: Routledge.

diSessa, A. A. (1993). Towards an epistemology of physics. *Cognition and Instruction, 10*(2&3), 105–225. https://doi.org/10.1080/07370008.1985.9649008.

Driver, R., & Easley, J. (1978). Pupils and paradigms: A review of literature related to concept development in adolescent science students. *Studies in Science Education, 5,* 61–84. https://doi.org/10.1080/03057267808559857.

Driver, R., & Erickson, G. (1983). Theories-in-action: Some theoretical and empirical issues in the study of students' conceptual frameworks in science. *Studies in Science Education, 10,* 37–60. https://doi.org/10.1080/03057267808559857.

Driver, R., & Oldham, V. (1986). A constructivist approach to curriculum development in science. *Studies in Science Education, 13,* 105–122. https://doi.org/10.1080/03057268608559933.

Driver, R., Leach, J., Millar, R., & Scott, P. (1996). *Young people's images of science*. Buckingham: Open University Press.

Duit, R. (1991). On the role of analogies and metaphors in learning science. *Science Education, 75*(6), 649–672. https://doi.org/10.1002/sce.3730750606.

Duit, R., & Treagust, D. F. (1995). Students' conceptions and constructivist teaching approaches. In B. J. Fraser & H. J. Walberg (Eds.), *Improving science education* (pp. 46–69). Chicago, IL: The National Society for the Study of Education.

Erman, E. (2017). Factors contributing to students' misconceptions in learning covalent bonds. *Journal of Research in Science Teaching, 54*(4), 520–537. https://doi.org/10.1002/tea.21375.

Fetherstonhaugh, T., & Treagust, D. F. (1992). Students' understanding of light and its properties: Teaching to engender conceptual change. *Science Education, 76*(6), 653–672. https://doi.org/10.1002/sce.3730760606.

Gabel, D. L. (1989). Let us go back to nature study. *Journal of Chemical Education, 66*(9), 727–729. https://doi.org/10.1021/ed066p727.

Gilbert, J. K., & Watts, D. M. (1983). Concepts, misconceptions and alternative conceptions: Changing perspectives in science education. *Studies in Science Education, 10,* 61–98. https://doi.org/10.1080/03057268308559905.

Griffard, P. B., & Wandersee, J. H. (2001). The two-tier instrument on photosynthesis: What does it diagnose? *International Journal of Science Education, 23*(10), 1039–1052. https://doi.org/10.1080/09500690110038549.

Hasan, S., Bagayoko, D., & Kelley, E. L. (1999). Misconceptions and the Certainty of Response Index (CRI). *Physics Education, 34*(5), 294–299. https://doi.org/10.1088/0031-9120/34/5/304.

Hewson, P. J. (1981). A conceptual change approach to learning science. *European Journal of Science Education, 3*(4), 383–396. https://doi.org/10.1080/0140528810304004.

Johnstone, A. H. (2000). Teaching of chemistry—Logical or psychological? *Chemistry Education: Research and Practice in Europe, 1*(1), 9–15. https://doi.org/10.1039/A9RP90001B.

Lin, J.-W. (2016). Development and evaluation of the diagnostic power for a computer-based two-tier assessment. *Journal of Science Education and Technology, 25*(3), 497–511. https://doi.org/10.1007/s10956-016-9609-5.

Loh, A. S. L., Subramaniam, R., & Tan, K. C. D. (2014). Exploring students' understanding of electrochemical cells using an enhanced two-tier diagnostic instrument. *Research in Science & Technological Education, 32*(2), 229–250. https://doi.org/10.1080/02635143.2014.916669.

Osborne, R. J., Bell, B. F., & Gilbert, J. K. (1983). Science teaching and children's views of the world. *European Journal of Science Education, 5*(1), 1–14. https://doi.org/10.1080/0140528830050101.

Palmer, D. H. (1999). Exploring the link between students' scientific and nonscientific conceptions. *Science Education, 83*(6), 639–653. https://doi.org/10.1002/(SICI)1098-237X(199911)83:6%3c639:AID-SCE1%3e3.0.CO;2-O.

Peterson, R. F., Treagust, D. F., & Garnett, P. (1989). Development and application of a diagnostic instrument to evaluate grade-11 and -12 students' concepts of covalent bonding and structure following a course of instruction. *Journal of Research in Science Teaching, 26*(4), 301–314. https://doi.org/10.1002/tea.3660260404.

Ryder, J., & Leach, J. (2000). Interpreting experimental data: The views of upper secondary school and university science students. *International Journal of Science Education, 22*(10), 1069–1084. https://doi.org/10.1080/095006900429448.

Solomon, J. (1993). The social construction of children's scientific knowledge. In P. J. Black & A. M. Lucas (Eds.), *Children's informal ideas in science* (pp. 85–101). London: Routledge.

Sreenivasulu, B., & Subramaniam, R. (2013). University students' understanding of chemical thermodynamics. *International Journal of Science Education, 35*(4), 601–635. https://doi.org/10.1080/09500693.2012.683460.

Taber, K. S. (1999). Ideas about ionisation energy: A diagnostic instrument. *School Science Review, 81*(295), 97–104.

Taber, K. S. (2000). Chemistry lessons for universities?: A review of constructivist ideas. *University Chemistry Education, 4*(2), 63–72.

Taber, K. S. (2003). Understanding ionisation energy: Physical, chemical and alternative conceptions. *Chemistry Education Research and Practice, 4*(2), 149–169.

Taber, K. S. (2006). Beyond constructivism: The progressive research programme into learning science. *Studies in Science Education, 42*(1), 125–184. https://doi.org/10.1080/03057260608560222.

Taber, K. S. (2013). *Modelling learners and learning in science education: Developing representations of concepts, conceptual structure and conceptual change to inform teaching and research*. Dordrecht: Springer.

Taber, K. S. (2014). *Student thinking and learning in science: Perspectives on the nature and development of learners' ideas*. New York: Routledge.

Taber, K. S., & Tan, K. C. D. (2007). Exploring learners' conceptual resources: Singapore A level students' explanations in the topic of ionisation energy. *International Journal of Science and Mathematics Education, 5*(3), 375–392. https://doi.org/10.1007/s10763-006-9044-9.

Tan, K. C. D., Goh, N. K., Chia, L. S., & Treagust, D. F. (2002). Development and application of a two-tier diagnostic instrument to assess high school students' understanding of inorganic chemistry qualitative analysis. *Journal of Research in Science Teaching, 39*(4), 283–301. https://doi.org/10.1002/tea.10023.

Tan, K. C. D., Taber, K. S., Goh, N. K., & Chia, L. S. (2005). The ionisation energy diagnostic instrument: A two tier multiple-choice instrument to determine high school students' understanding of ionisation energy. *Chemistry Education Research and Practice, 6*(4), 180–197. https://doi.org/10.1039/B5RP90009C.

Tan, K. C. D., Taber, K. S., Liew, Y. Q., & Teo, K. L. A. (in press). The web-based ionisation energy diagnostic instrument: Exploiting the affordances of technology. *Chemistry Education Research and Practice*. https://doi.org/10.1039/C8RP00215K.

Tan, K. C. D., & Treagust, D. F. (1999). Evaluating students' understanding of chemical bonding. *School Science Review, 81*(294), 75–83.

Treagust, D. F. (1995). Diagnostic assessment of students' science knowledge. In S. M. Glynn & R. Duit (Eds.), *Learning science in the schools: Research reforming practice* (pp. 327–346). Mahwah, NJ: Lawrence Erlbaum Associates.

Treagust, D. F., Duit, R., & Fraser, B. J. (1996). Overview: Research on students' preinstructional conceptions—The driving force for improving teaching and learning in science and mathematics. In D. F. Treagust, R. Duit, & B. J. Fraser (Eds.), *Improving teaching and learning in science and mathematics* (pp. 1–14). New York: Teachers College Press.

Voska, K. W., & Heikkinen, H. W. (2000). Identification and analysis of student conceptions used to solve chemical equilibrium problems. *Journal of Research in Science Teaching, 37*(2), 160–176. https://doi.org/10.1002/(SICI)1098-2736(200002)37:2%3c160:AID-TEA5%3e3.0.CO;2-M.

Yan, Y. K., & Subramaniam, R. (2018). Using a multi-tier diagnostic test to explore the nature of students' alternative conceptions on reaction kinetics. *Chemical Education Research and Practice, 19*(1), 213–226. https://doi.org/10.1039/C7RP00143F.

Ye, L., & Lewis, S. E. (2014). Looking for links: Examining student responses in creative exercises for evidence of linking chemistry concepts. *Chemistry Education Research and Practice, 15*(4), 576–586. https://doi.org/10.1039/C4RP00086B.

Chapter 11
A Longitudinal View of Students' Perspectives on Their Professional and Career Development, Through Optional *Business Skills for Chemists* Modules, During Their Chemistry Degree Programme

Samantha Louise Pugh

Abstract Employers regularly cite a lack of commercial awareness and other transferable skills in new graduates. To address this issue, we developed a suite of employability-focused modules (or courses) under the umbrella of *Business Skills for Chemists*. The modules were, and continue to be, optional for all Chemistry undergraduates, with one 10-credit module (of 120 credits per year) for each undergraduate year of study. A context-based learning approach and group work was taken in each case, to introduce students to a wide range of industrially focused experiences. A retrospective longitudinal qualitative study of three students who took all three modules during their degree was undertaken to better understand students' experience of the modules, and the impact on their career decision-making. This research was undertaken to gain greater insight than the regular student feedback obtained at the end of each module. During the modules and at the end of the degree, students identified that the modules had helped them to develop a wide range of skills and capabilities. Reflection is an essential component of the learning experience, ensuring that the students not only experience a skills-rich curriculum, but also have the ability to reflect upon and derive benefit from their experiences. The modules had also been prominent in their career decision-making, by introducing the students to a wide range of career options for Chemists, through the curriculum.

Introduction

The development of students' transferable skills is essential during a degree programme. This is clearly articulated in the Quality Assurance Agency benchmark statements for Chemistry (QAA, 2014) and in the Royal Society of Chemistry's

S. L. Pugh (✉)
University of Leeds, Leeds, UK
e-mail: s.l.pugh@leeds.ac.uk

M. Schultz et al. (eds.), *Research and Practice in Chemistry Education*,
https://doi.org/10.1007/978-981-13-6998-8_11

accreditation framework (RSC, 2017). Of relevance to this intervention, the QAA benchmark statements (QAA, 2014) explicitly refer to:

- Communication skills, covering both written and oral communication with a variety of audiences;
- Information location and retrieval skills, and the ability to assess the quality of information accessed;
- Basic interpersonal skills, relating to the ability to interact with other people and to engage in team working;
- Time management and organisational skills;
- Other relevant professional skills such as business awareness.

Similarly, the RSC accreditation framework refers to the development of professional skills (RSC, 2017):

- Communication skills;
- Scientific writing, data presentation, referencing literature;
- Ethical responsibilities;
- Sourcing of information;
- Team working;
- Time management and organisational skills.

Traditionally, the solution was to include a 'skills module' within the degree programme; however, these were often ineffective because they lacked chemistry context, and therefore students did not recognise them as part of their core learning of the subject (Tomlinson, 2012). Similarly, Hill, Overton, Thompson, Kitson, and Coppo (2018) noted that students are not always aware of the skills that they are developing through the curriculum. Additionally, a study by Hanson and Overton (2010) examined the skills that were typically developed in a chemistry degree, contrasted with the skills that employers seek. There was a mismatch, particularly with respect to development of transferable skills. A study by Galloway (2017) found that students particularly valued developing skills such as teamwork, problem solving, and organisational or time management, among other chemistry-related skills.

One solution is to embed context- or problem-based learning into the degree programme, thus providing students with a more authentic learning experience. Authentic learning is a term that is used to describe learning by doing (Pearce, 2016) that takes place in a realistic, or simulated real-world context. Four components of authentic learning are (Rule, 2006):

1. Problems that mimic the work of professionals;
2. An aspect of inquiry or open-ended learning;
3. Students engage in a community of learning;
4. The students direct the learning.

An authentic learning pedagogy was used in the development of the courses that are referred to in this chapter. This approach was adopted to give students experience of realistic, workplace tasks and to hopefully increase engagement by means of intrinsic motivation, where autonomy and purpose would be key drivers (Herzberg, 1966; Pink, 2010).

The value of context-based and problem-based learning is well documented (Belt, Leisvik, Hyde, & Overton, 2005; Seery, 2015). An article by Overton and Randles (2015) cites many examples of where PBL is used in chemistry education successfully.

We developed a suite of three 10-credit context-based modules (or courses) under the umbrella of *Business skills for Chemists*; one for each 120-credit year of a three-year UK undergraduate Chemistry degree. The aims of the modules were to increase students' commercial awareness and business acumen (Wilkinson & Aspinall, 2007), provide students with an opportunity to develop the transferable or professional skills specified by the QAA and the RSC, and provide an opportunity to reflect on their experiences. All three of the modules were set in the context of the chemical industry, each having a different focus. All of the modules included guest speakers with a range of career profiles within the chemical sector. The skills that were intended to be developed through the modules were:

- Communication
- Information retrieval
- Teamwork
- Persuasive skills
- Commercial awareness.

The first-year module focused on employment in small to medium enterprises (SMEs), an often overlooked sector for students seeking employment (Pugh, 2017). The SME case study formed part of a career planning module, and provided students with the opportunity to select an existing company, then enact the role of employees of that company whilst completing an extended activity considering the most promising commercial opportunities for their chosen company. Students worked in allocated teams, made decisions about the opportunities over a period of weeks, and then presented their preferred opportunity, as a group, to their line manger, enacted by the module leader.

The second-year module focused on new product development in a chemical-related business, taking a product from concept through to market (Pugh, 2014). This module introduces students through industry-led workshops to intellectual property, marketing, scale up, legislation, project management and other commercial considerations. Students are assigned groups, given a project brief and meet on a weekly basis to work on their project. Weekly sessions are scheduled, although students are expected to arrange meetings with each other outside of the scheduled slots. Each group presents their proposal to the 'board of directors,' enacted by the module team, as a pitch at the end of the module. They also prepare a group portfolio of their work.

The third-year module was developed in collaboration with the university's student business start-up unit, and is entrepreneurial in nature. Students are asked to find a piece of chemistry research that has potential for commercialisation or societal benefit and relates to the Royal Society of Chemistry's priority areas (RSC, n.d.), and develop a business case. At the end of the module, the students present a pitch to 'a team of investors,' enacted by the module team and external panel members, and also submit a business plan.

The modules were designed to be progressively more challenging. The first was very prescriptive, based on existing businesses and information was provided to students. In the second, the students were assigned a project brief for a new product development in a specified sector but they had to find their own information. In the third, the students had the freedom to choose their own area of research and design their own product. This increased level of autonomy was commensurate with their level of study and experience.

Throughout all of the modules, students work in teams. Their assessment consists of a group report, a group presentation, and an individual reflection on their experience. The approach to reflection followed the Kolb learning cycle which progresses from having an experience, to reflecting on that experience, conceptualising what has been learned, then putting it into practice in the future (Stice, 1987). Students are introduced to reflection in a taught session before the end of the module in each case. There will be repetition for students who have taken more than one module, but as there are no prerequisites, it cannot be assumed that all students have previously learned about reflection in earlier years.

Reflective essays were included because it is not enough for students to just participate in skills-rich activities. They need to have the opportunity to critically reflect upon their experience in order to derive benefit from it (Harvey, 2005). Reflective practice is used widely for self-development in medical education (Riley-Douchet, 1997), but less so in the physical sciences. The author believes that the benefits of reflection are two-fold for students. Firstly, they gain experience in articulating their experiences and the skills that they have developed, which is a useful skill for future interviews. Secondly, it encourages self-evaluation and identification of strengths and weaknesses, which will be useful if students are to become lifelong learners, capable of identifying their own development needs.

Student feedback on the modules has generally been positive in all cases. The majority of students enjoy the modules and particularly that they have an opportunity to shape their own learning and have some autonomy in their work. The students describe working in a team as a mixed blessing. Some students have a really excellent team where everyone works well together; others do not. Overall, even when students did not enjoy the experience, they valued the skills that it allowed them to develop for the future. Through their reflections, they often state that these modules are the only place where they get to develop such skills. The feedback for the third-year module has 100% satisfaction consistently. However, as these students have normally taken the second-year module, they have a much better understanding of what is required and are used to this way of working.

Anecdotally, there is also evidence that the modules in years one and two have helped students to secure a year in industry, although it would be hard to demonstrate that this is causal, given that they are optional modules. Through the reflective essays, many of the students talk about how the modules have helped with their career decision-making.

Typically, 33–40% of the cohort of B.Sc. Chemistry students takes the modules in years one and two, and around 20% in year three. The drop off can be explained to some extent by the fact that some of the students from year three are spending a year in industry as their third year instead.

Methodology

This paper focuses on the way that students articulate and derive benefit from their learning experiences. The research received ethical approval from the University of Leeds (MEEC-13-017, April 2018).

The qualitative research involved two aspects: a focus group, and analysis of historic reflective essays. Three students who had taken all three modules took part in a focus group at the end of their degree programme to discuss their experiences. The students also gave the researcher permission to review their original reflective essays from each of the three modules. After the focus group had taken place, each of the students' submissions were reviewed in turn, and in chronological order to make sense of their perspectives at different points during their degree. In each case, the reflective essay was written at the end of first semester in each year of study. A retrospective longitudinal approach was chosen because students can change the way that they narrate their decision-making over time, as described by Cleaves (2011) and so this approach also examined if the students changed their minds or narrated their experiences differently over time. A focus group was chosen rather than interview as individual perspectives had already been captured through the reflective essays and the author was also interested in the way that the students interacted with each other during the discussion. There was evidence during the discussion that the focus group helped the students to make further sense of their experiences through the group discussion.

The students have been given the pseudonyms Lara, Mark and Gina. All three students were just about to graduate and were awaiting results from their undergraduate BSc Chemistry programme when the focus group took place. The undergraduate Chemistry degree consists of a common core, with organic, inorganic and physical chemistry, mostly taught in a traditional lecture-based format with unseen written exams and weekly homework as the main forms of assessment. The programme also has a large practical element through each of the three years (around 25% of the programme). In each year, students have a small amount of credit (typically 10–20%) for optional modules. These options are normally taken within Chemistry although broader subjects are permitted.

Each of the case studies was considered in turn, taking into account information gleaned from their reflective essays and their responses during the focus group discussion. Thematic, rather than chronological, analysis of the essays and the transcript was based on the following themes:

1. Why did they choose the 'Business Skills for Chemists' pathway?
2. What were the benefits?
3. What were the challenges?
4. What was the impact on their future career decision-making?

Results and Discussion of Case Studies

Three case studies are presented.

Case Study 1: Lara

Lara studied B.Sc. Medicinal Chemistry. The Medicinal Chemistry programme follows the common core of the Chemistry programme, but with some additional specialist modules. During the year one module, from her essay, Lara was wholly positive about group work and appreciated the value of working in a team. All members of her team were committed and worked well together. During the course of the module, Lara was "put off working for an SME… due to high responsibility and high visibility in a small business." However, she recognised that, "You would need to have good team work skills" and, "your ideas would be heard by everyone, which could be both an advantage and a disadvantage."

There was definitely a change of tone after the second-year module. Lara's opening statement in her essay was, "Group work can be a struggle." She found the team frustrating at times but she assumed a leadership role to get the team back on track. This resolved the initial issues. Lara stated that she wants to be a drug designer, and related this to the good presentation skills she was developing through the module. She said that the module, coupled to some of her co-curricular activities, helped to increase her confidence, and that she would need confidence as a drug designer, "if I think I have made a particularly important discovery, I will have to possess the confidence to present the results." There was also evidence of Lara envisaging herself as a drug designer in the future.

Overall, Lara highlighted that the module, "Helped me to acknowledge my strengths and weaknesses and given me chance to improve upon them."

After the completing the third-year module, there were some similarities with the second-year experience, in that the group work proved a struggle, especially at the beginning of the project work. However, in this reflective essay, Lara focused much more on the wider benefits of the module, for example, "It has allowed me to become more competent in an assessment centre situation. Having done a mock presentation before… I felt much more comfortable presenting."

The third-year module presented additional challenges, in that the students had to generate their own idea, rather than being given a product brief, "We found it really difficult to knuckle down to an initial idea." Once again, Lara took on a leadership role within the team, "I decided to solve the problem by delving into my own interests [in healthcare, and related to co-curricular activities]… and the rest of the group were happy to get started." This was then supported by research to find the appropriate technology. The initial aims of the modules were not to develop research skills, but this has been an unintended and unexpected benefit, with several students highlighting how they have improved their research skills through the modules, and for a real purpose, rather than just for the sake of an assignment. There was also clear evidence of commercial awareness being developed through the process, "We would know for sure that our product would be in demand, which is vitally important for business." There was also evidence of the learning experience being authentic in that Lara envisaged herself in a professional role, "So that is how we became a healthcare company." There was also evidence that the module was allowing students to develop higher order learning in accordance with Bloom's Taxonomy (Bloom, 1956), particularly creating knowledge, "I enjoyed creating and developing our idea."

Teamwork was discussed further, with Lara recognising that each member of the team can make a big difference, and she found this really motivating. Lara identified that patience is an area that she needs to develop, although she hadn't originally realised this. She felt let down by other team members, and this was set against a backdrop of her own life being very busy with various commitments, but she still found the time to do the work. However, despite frustrations, she always sought to encourage others.

By this time, Lara had changed her career aspirations, and stated that she wanted to become a clinical scientist, and made no reference to the earlier ambition of being a drug designer.

At the time of the focus group, Lara said that she originally chose the modules because of the nature of the assessment, "I liked… the presentation, the self reflection and the portfolio was a better way than the stresses of exams for me." She thought it was good to find out what it is like in industry, and to develop presentational skills for when they go into industry. The module also opened her eyes to career opportunities for Chemists.

Lara said that she opted to take the second-year module because she had enjoyed the first-year module and thought it would be similar, and the content was quite interesting, "It's just nice to see the reality of what goes on." Lara also liked that the module gave her the opportunity to talk to people from industry, "Because we don't get the opportunity in really any other modules to do that." This wasn't something that she had mentioned during any of the reflective essays. She also highlighted the importance of developing presentational or teamwork skills, which they don't do in other modules. "It just throws in a whole different new skills set that's kind of refreshing."

Lara mentioned that she liked, "Making sure that everyone is organised." This theme cropped up in her second-year and third-year reflections, and also when she talked about her co-curricular activities. Lara also identified learning new skills in the field of marketing, such as SWOT (strengths, weaknesses, opportunities and threats) and PESTLE (political, economic, social, technological, legal and environmental) analysis. Again, Lara demonstrated the development of her commercial awareness, "You've already got the demand, and you can shape it that way. You know it works and people are going to want to buy it if it works."

In the focus group, Lara said that she was quite open-minded about careers, although she had found the role of 'clinical scientist' in the National Health Service (NHS) that interested her. However, as she was just finishing university, "I'm not sure I'm 100% ready to just be thrown into a lab situation. So I've been thinking about teaching English in Spain." She is also considering going into teaching. It's an interesting point that maybe students aren't always ready to commit to a career path immediately after finishing their studies, and they want the time and space to make their decisions. She highlighted also that she liked being around people, so that would be a consideration in a career choice. She added that in the past she had shadowed people who worked in careers she had considered, but this had tended to put her off. One of the issues highlighted was that the options for a chemistry graduate are really broad, so it can be difficult to home in on what they want to do. Her view was that this was generally how most of her friends felt about careers too. "I think it's definitely more unusual that some knows 100% that I'm going to apply for this, I'm going to do this kind of thing."

Lara's closing comment was "All the modules… they've just provided a really good range of things to do and it's been really interesting for me."

Case Study 2: Mark

Mark studied B.Sc. Chemistry. Mark identified in his essay that during the first-year module, his project team didn't hold enough group meetings, so everything was rushed at the end and not well summarised. He also felt that he wasn't well prepared for the presentation and that he would plan ahead better in the future. Through the module, Mark recognised that he lacked experience to put on his CV, so said that he would seek out more opportunities in the future. We will return to this point later.

Mark identified at the end of the module that he now had a greater understanding of the value of deadlines and time management, "if we'd started out earlier and set clearer goals, and then set up the meetings to summarise everyone's work… so none of the group feels panicked of under pressure [for the presentation]."

Mark highlighted that the skills he'd developed were, "More of a change in attitude to group projects and how I can apply my skills to make me a better team player."

After the second-year module, Mark set the scene in his essay with his previous experiences of workshops throughout his education, which had made him apprehensive. He was used to, "Improvising on the spot so this module was vastly different as I had to prepare weeks in advance learning knowledge about my topic." Mark chose to take responsibility for the technical aspects of the project, "I thrive on telling facts and figures as well as mechanisms."

Again, Mark identified that the module was an opportunity for authentic learning, "This type of work, which is closest to an actual job." Mark then went onto say that often it is not the skill set, but the mind set that is important for success in a project such as this. Also, Mark related this to a future career, where the depending on the nature of the role, he would need to learn to adapt and evolve his skillset.

By the third-year essay, Mark identified that his attitude had improved significantly, and said, "I took the module based on taking the previous module, and to further increase my knowledge of business as a whole." He also hoped to increase his confidence and to improve commercial awareness; this latter point was evident in a number of comments that he made, for example, "There's a fine balancing act to make an idea financially viable," and "This helped turn it into a manageable debt which was done by introducing capital to pay for all the initial stages." In general, these are not topics that a chemist would consider during their undergraduate programme. Mark said that he struggled with the initial research but grew in confidence once he'd read into things more deeply. He got better at organising his priorities and improved his own writing skills by working as part of a group. He said that the group worked well together. Mark did not make any reference to his career decision-making.

At the end of Mark's degree, in the focus group he said that he hadn't put any great thought into taking the business skills modules. He didn't know what he wanted to do at that point and thought, "business skills, handy." There was also the added benefit that the module didn't have an exam. The fact that the modules didn't have exams was part of the reason for choosing the business skills options throughout the degree. "I definitely needed one less exam, just so then I can cope." The opportunity to learn about the chemical industry was also cited as a reason for choosing the second-year module, "maybe I wanted to be more in-depth into what chemistry in industry is like and what basically is the market for it."

Mark said that he'd never been apprehensive about working in a group and tends to "edge my way in" and take on a leadership role. In second year, he said he wanted to relax a bit and as he was with a group of friends, they all managed to split the work evenly. This raises an interesting question about whether students should be assigned groups (as in this case) or be allowed to choose to work with friends. There are pros and cons to either, but we have tended to allocate groups as this more closely replicates the workplace. Mark felt that the modules have strengthened his skills with working with people and to behave more professionally. He also learned how to network.

During the focus group discussion, Mark said that he learned which areas of business he liked and disliked, "I wanted nothing to do with marketing pretty much." He also said this has influenced his career thinking, "I don't want to deal with like marketing, advertising and all of that. I'd rather be in with the science and know what I know. I just want to be the person that's in the labs." Throughout the conversation, Mark reiterated this point several times. He also said, "I might want to be the person that explains the science to the business person, but that's probably as for as the business for me would interact." It is really useful for a student to be exposed to different aspects of business at this stage in their learning—they can get a flavour for what they do and do not like. In each of the modules, Mark opted to take "the science role." Mark also took on the finance role, "I did the finance one, my god, that was annoying… Finance is not my thing." Lara added to the focus group discussion by referring to another team member, Joe. "He made up for your finance, and then you helped with the science side of it." There was a sense that whilst the students assigned themselves specific roles, there was a lot of interaction between the different functions throughout the project; in fact, this was necessary for the project to be delivered successfully as all the roles were interrelated.

Mark demonstrated commercial awareness through comments that were made during the discussion, for example: "Ah this thing is as cheap as chips now, where you can probably do this as a feasible business idea." And, "It was a decent make up and we could make a profit within the first year." There was also evidence of authentic learning taking place during the third-year module, "I wish we could make this now" indicating that the process had seemed real and immersive, and that an actual physical product was, in his eyes, a viable outcome for the project.

In terms of career plans, Mark recalled that when he first went to University he was thinking of a career in making anti-cancer complexes, because there is a big demand for healthcare drugs, but the complexity of the science had put him off.

However, after completing his final-year research project, he is now wanting to go into a career in metallurgy, "Mainly just because of the project that I've done, like that was the most enjoyable thing of the whole degree for me… If there's a business for it, I'm going to find it." It is valuable for students to have a wide range of experiences during their degree programme; not just for enjoyment, but to help them to consider their possible career options. Also, it is not always an option for students to gain meaningful work experience during their studies, for a variety of reasons, as Mark explains, "I've only got the bare minimum work experience because I've always had to work to get my grades… so applying for a part-time job in between that as well as being there for my disabled parents as well, that's kind of hard." It is vital that the curriculum provides the opportunities that students will need to be successful post graduation. The co-curricular opportunities are not an option for everyone and so cannot be relied upon for wider skills development. Mark's final comments were, "I'm glad that I know the business side is not what I want to do, it's narrowed down what I actually want." An important message in this statement is that students don't necessarily need to enjoy something, or think that it is the career choice for them, for them to derive benefit from the experience. Although Mark didn't enjoy aspects of the project (i.e. finance), it did give him a valuable experience in being able to narrow down areas of work that did not interest him.

Case Study 3: Gina

Gina was also studying B.Sc. Medicinal Chemistry. Gina said in her essay that her first-year experience was a positive one. All of her group members were enthusiastic and organised. She said that working with strangers definitely developed her communication skills. Making group decisions was easy because, "We all got along well and had similar ideas." The module required a greater level of organisational skills than she had encountered previously, and needed the students to organise meetings outside of the scheduled classes in order to complete the work. The group made decisions quickly so didn't need to meet often, however, this led to an unexpected issue of needing to recap each time they met. A possible solution identified for future work by Gina would be to hold shorter but more regular meetings.

Gina particularly enjoyed the presentation, especially as she had not done many beforehand. Plenty of practice both alone and with the group, ahead of the assessment, meant that she felt confident delivering the presentation. "Practising as a group was vital in order to ensure smooth handovers and succinct presentation was delivered." Gina felt that overall the module had given her a good insight into what SMEs are and the challenges they face, and on a personal level increased her confidence in working in a group and giving presentations.

In her second essay, one of the key reasons why Gina chose to take the second-year module was to develop skills in organisation, group work, presenting, and communication. Additionally she wanted to gain an insight into how a new product is developed. At the time, she wished to work in pharmaceutical development.

Working in a group with people she did not know did not faze Gina as she had got used to working and interacting with strangers since moving to university. However, she did say, "This was probably the most complex and demanding task I have had to complete when working with a group of unknown people." She felt that the whole group got on well and were equally motivated to complete the project. One of the biggest challenges they found as a group was finding a time when they were all free to meet, due to a heavy timetable and other co-curricular and work commitments. This is something that course designers need to consider when projects are introduced that require students to work together outside of scheduled classes. To address this issue, at least partially, the students divided up the project requirements and delegated them to different members of the team. They also set up an online group chat to enable exchange of ideas remotely. This was an effective way of progressing the project, but not without its problems, "In hindsight, I feel that we shouldn't have delegated each section of the project so early… it meant that each person only got to look in-depth at one area of the project. I feel I would have learned more from the project if I had the opportunity to research in most, if not all, areas… It may have been easier for each of us to complete our own area of the project if we had greater understanding of what the others were doing." This point illustrates the difference between a team and a group. I team is more integrated and creates a more synergistic effect.

Gina felt that the presentation went well, "I clearly and confidently communicated our marketing strategy to the board." However, she noted that the group didn't deal with the questions well, with everyone trying to answer all questions, "and this seemed quite unprofessional." She reflected that the group should have also devised a strategy for dealing with the questions.

After the module, Gina's understanding of how to create a viable product had grown. Additionally, the module had an impact on her career thinking, "I had never studied marketing before, but it is something that I am now keen to learn more about." Gina also makes reference to her future self, "As a pharmaceutical development scientist, I may have to work on projects similar to this," which indicates an authentic learning experience, and one that is helping students to make career choices. The only downside that Gina identified was not being able to gain in-depth knowledge in all aspects of the product development, but that is the nature of a team project, and more closely resembles the workplace.

By third year, Gina decided to take the module because she had enjoyed the second-year module. She identified that she wanted to develop teamwork, presentation, and organisational skills. The second-year module had sparked an interest in marketing for Gina, and she hoped to learn more about this area of business.

She also liked that they had more creative control over the project, compared to year two. However whilst the creative control was attractive, it proved very challenging during the early weeks of the project, "Choosing from such a vast range of ideas proved more difficult than expected. Some group members were negative towards every idea yet not coming up with any themselves." At this point, Gina took leadership of the group and suggested everyone come up with one idea each for the next meeting, to pitch to the rest of the group. The groups were not given any instructions about how they should structure themselves, or how they should make decisions. It is interesting to observe that in most groups, a leader emerges at some point, particularly if there are challenges.

Again, the work was delegated to different team members, and Gina set up a group chat, "Because I knew that this would be an effective way of working." However, they held regular meetings to ensure that everyone was involved with all aspects of the project. It was good to see that Gina had reflected and learned from her previous experience of a similar project. In the previous year, Gina felt that everything was rushed at the end, so this time they created "a rough plan of when things needed to be completed by." It was good to see that they put their project management training from second year into practice. Even more pleasing was that this skill is also being used for other activities, "This is a method I use frequently to manage my own time, but the first time I had applied it to a group."

Overall, Gina felt she had developed her organisational skills, time management and communication skills in a team. Interestingly, Gina did not identify time management as a skill to develop at the start of the module. She also noted that she had reflected on previous experiences and applied the learning to this module, "I was also able to reflect on what was successful in the second year and apply this to the project this year. As a result, the project was more professional than work I had completed previously."

In her essay, Gina summarised her career planning by saying that she was considering starting her own business in the future and felt that "I have gained an excellent insight into what I would need to do in order to make this a reality."

At the end of the degree, during the focus group Gina said that she chose the first-year module because she was planning to do a placement year and, "It would help get your CV done" but also found the module useful. She felt that it had opened her eyes to career opportunities, and especially to SMEs, "Because I'd never really thought about those sorts of companies before." She found, "the biggest challenges were working with others, but enjoyed that they got to do group work." Also, she emphasised that in other modules, there is no group work, "Actually [group work] is quite a useful skill when you go into a job." The biggest challenge for the group work was finding time to communicate with each other.

Gina also talked again about learning about advertising and marketing, "It was just a bit more variation" and gave her an alternative career option, "I don't want to work in a lab, but I've learnt that I want to work with people." She also identified that she had an interest in business (and not necessarily chemistry business). This was an interesting contrast to Mark who definitely did not want to go into business. She noted that the module gave them the opportunity to communicate science to a different audience, "Explaining the science in like a simplified way."

She didn't like being assigned a project brief in second year so much, as she didn't find the topic engaging, "Effluent water treatment, and it was, no offence, so boring." However, the group recognised that they needed to think outside the box to make it more interesting. She really enjoyed being able to pick their own project in third year because they could, "Personalise it a bit more." In hindsight Gina did not initially mention the challenge that the group had faced in identifying a product in third year. After further discussion about why the second-year module is more structured, she said "I totally got that because then this year was like, Oh my God, we've just got to think of something out of thin air."

Gina said that she enjoyed doing the marketing aspects, and that she volunteered to take on this role in the third year. She liked to find engaging ways to present to people; also using social media to communicate to a wider audience. A student creating a social media profile for a fictional product demonstrates the level of engagement with the task.

Gina reaffirmed that she did not want a career that involved working in a lab, but would be interested in science communication, "where I can do outreach, and that sort of thing." She plans to apply for the Wellcome Trust internship programme, but is in no rush, as she hasn't had time to fully think about her future career or make any applications, "I work on the weekends and I'm at Uni in the week so…" She felt that there was no rush, and happy to waitress for a while once she had graduated and had more time to think about what she wanted. All the participants thought that this was common among the BSc students but the Masters students are more likely to have career plans confirmed.

On reflection, Gina felt that the modules had, "Opened up our eyes to other careers without really having to go out of your way to look for them… and then sparked an interest and you can go and look yourself a bit more." She went on to say that this didn't generally happen with any of their other modules, "It's like hard to imagine how other than working in a lab how these things could apply elsewhere, but this is the only module that actually says there is other things as well that aren't so science."

Outcomes of Intervention

The case studies provided a detailed perspective on how the students narrate their own experiences and derive benefit from those experiences. Here, I will reflect on the four themes that were outlined at the start of the methodology, namely:

1. Why did they choose the 'Business Skills for Chemists' pathway?
2. What were the benefits?
3. What were the challenges?
4. What was the impact on their future career decision-making?

The students did not appear to give a great deal of thought to why they had opted to take the first-year module. The main drivers were two-fold. Firstly, they all thought that the opportunity to create a curriculum vitae (CV) and learn more about the chemical industry would be useful. Secondly, students were attracted to the module because it did not have a formal examination, whereas all of the other chemistry modules (except laboratories) did. When it came to the second year, again, the absence of a formal examination was attractive. However, the students also identified that 'business skills' would be useful for their career futures. Additionally, they had enjoyed the experience of working as a group in year one, so the module felt like less of an unknown entity. Moving onto year three, the motives were very similar to year two. The students were also keen to have the opportunity to develop their own ideas, although that brought some challenges too.

In terms of benefits, the participants self-identified the following skills and competency development through participation in the modules: Teamwork, organisational skills, presentation skills, time management, professionalism, communication, decision-making, research skills, planning, confidence, lateral thinking, business acumen, networking, change in attitude, reflection, problem solving and leadership. This list went far beyond the initial intentions of the module, described as: Communication, information retrieval, teamwork, persuasive skills and commercial awareness. For many of the skills, students stated that these modules were the only place throughout their degree where they had an opportunity to develop these skills.

The challenges that the students faced throughout the modules were largely due to working in a team. The modules are the only place during their degree where they formally work in a team, although there are examples of co-operation in tutorials and laboratories. Understanding team dynamics and how to work with others was a big challenge, with the most challenging time being during year two. Due to their busy timetable and other commitments such as part-time work, finding a time to meet was also a challenge. This needs to be considered if courses are designed that rely on unscheduled group meetings. Another challenge that all of the students faced was regarding decision-making. The majority of their studies have been tutor-directed, with a focus on traditional didactic lectures. These modules needed the students to make their own choices throughout and direct their own learning throughout, with the level of autonomy increasing through the years. Students found this challenging, however, it is a much more authentic experience and good preparation for the workplace.

Throughout the reflective essays, and then during the focus group, all students referred to their career decision-making. Generally, students started out with some career aspirations but they were often generic or vague. As the students progressed through their programme, and as seen through the modules, students had a clearer picture of what they did and, just as importantly, did not want to pursue career-wise. It is valuable to provide students with the opportunity to explore possible career futures through the curriculum. This addresses the issues with expecting students to take a proactive approach, such as visiting a careers service. Through curricular learning, students are automatically exposed to career possibilities. It is really important, especially for students that have decided they do not want to work in a lab environment, that there are other related career options available to them (Mellors-Bourne, Connor, & Jackson, 2011).

As the modules progressed, the students became increasingly reflective and reflexive; that is, not only were they reflecting on their experiences, they were also putting what they had learned into practice for future work. There was evidence that the students drew upon their previous experience and applied the learning to their current work. Reflective practice is a useful life skill and will help students to become lifelong learners, by giving them the skills to identify their own developmental needs in the future.

Students demonstrated through the narratives that they had experienced authentic learning and that they had collective ownership of their projects. Ownership and autonomy are known to support intrinsic motivation, rather than relying on external motivators such as the lure of high grades. Once students leave education and enter the workplace, intrinsic motivation will have longer-term benefits as it can lead to greater job satisfaction and, ultimately, success (Pink, 2010).

Conclusions and Implications for Practice

There has been extensive evidence from the literature that context- and problem-based learning is a very valuable way for students to experience authentic learning and improve outcomes. This particular intervention adds to that body of evidence. Exposing students to different career profiles through their learning, particularly pathways such as marketing, finance and intellectual property, is invaluable in terms of ensuring that all students have the opportunity to consider their future career options, both within and beyond chemistry. There was evidence that the students found that the modules helped them to consider career options that they both did and did not want to pursue in the future.

A possible future development to address issues with team dynamics could be to include more discussion with the students about team dynamics and the different roles that members of a team may adopt. A useful framework is the Belbin Team Role Descriptors (Belbin Associates UK, 2016). If students are able to self-identify their ways of working, and also share that with their team members, it may help to address some of the issues that the teams faced. This will also be a useful awareness to take forward throughout their careers.

An unexpected benefit of the modules was the wide range of skills that students self-identified. The most surprising to the author was that students identified that the modules were the main place in the curriculum where they developed their research skills. Whilst the modules clearly needed students to undertake research to find information, this was not a part of the modules by design. This research experience has lasting benefits for the remainder of the programme, and especially for the students' final-year research capstone project.

This chapter has taken a qualitative approach, following the experiences of three students that studied all three modules. An interesting and comparative piece of future work would be to examine the impact on those students who opted to take only one of the modules. This work could be undertaken by examining their reflective essays. However, student consent would need to be granted beforehand, which has been a limiting factor in this particular study.

Whilst building a skills-rich curriculum is important, it is insufficient if students are not provided with the opportunities and support to reflect upon their experiences. This process of reflection is where students make sense of their experiences and derive the most benefit from them. Through this reflective process, students were able to articulate the wide range of skills and competences that they had developed during the modules. There was also evidence of the students taking forward their learning from one year to the next. In this particular case, the modules are an optional part of the programme, but given the wide range of benefits that have been identified, there is a compelling argument that such interventions should be part of the compulsory core curriculum. This would also align with the 'professional skills' element of the QAA benchmark statements (QAA, 2014) and RSC accreditation framework, as articulated in the introduction to this chapter.

Acknowledgements The author wishes to acknowledge:
The National HE STEM Programme, the Higher Education Academy, The Royal Society of Chemistry and the University of Leeds for their financial support and role as critical friend in creating the resources;
Their co-workers in the development and delivery of the modules: Christopher Pask, Patrick McGowan, Stephen Maw, Caroline Williams, Christopher Hone, Ben Hetherington, Kairen Skelley and Richard Doyle, all at the University of Leeds, Tina Overton, formerly of the University of Hull, and Paul Taylor, formerly of the University of Warwick;
The Chemistry Industrial Advisory Board of the University of Leeds, and other external contributors to the modules.

References

Belbin Associates UK. (2016). Team role summary descriptions. https://www.belbin.com/media/2307/belbin-team-role-summary-descriptions.pdf. Accessed December 2018.

Belt, S., Leisvik, M. J., Hyde, A. J., & Overton, T. L. (2005). Using a context-based approach to undergraduate chemistry teaching—A case study for introductory physical chemistry. *Chemistry Education Research and Practice, 6,* 166–179.

Bloom, B. S. (1956). *Taxonomy of educational objectives. Book 1: Cognitive domain* (2nd ed.). New York: Addison-Wesley Longman Ltd. ISBN 10-0582280109.

Cleaves, A. (2011). The formation of science choices in secondary school. *International Journal of Science Education, 27,* 471–486.

Galloway, K. W. (2017). Undergraduate perceptions of value: Degree skills and career skills. *Chemistry Education Research and Practice, 18*(3), 435–440.

Hanson, S., & Overton, T. (2010). *Skills required by chemistry graduates and their development in degree programmes*. Hull: Higher Education Academy, UK Physical Sciences Centre. http://www.rsc.org/learn-chemistry/resources/business-skills-and-commercial-awareness-for-chemists/docs/skillsdoc1.pdf. Accessed December 2018.

Harvey, L. (2005). Embedding and integrating employability. *New Directions for Institutional Research, 128*, 13–28.

Herzberg, F. I. (1966). *Work and the nature of man*. Cleveland: World Publishing Company, chapter 6, OCoLC 654512681.

Hill, M. A., Overton, T. L., Thompson, C. D., Kitson, R. R. A., & Coppo, P. (2018). Undergraduate recognition of curriculum-related skill development and the skills employers are seeking. *Chemistry Education Research and Practice*. https://doi.org/10.1039/c8rp00105g.

Mellors-Bourne R., Connor H., & Jackson C. (2011). *STEM graduates in non-STEM jobs* (p. 299). Cambridge, UK: The Careers Research & Advisory Centre (CRAC).

Overton, T. L., & Randles, C. (2015). Beyond problem-based learning: using dynamic PBL in chemistry. *Chemistry Education Research and Practice, 16*(2), 251–259.

Pearce, S. (2016). Authentic learning: What, why and how? *E-Teaching: Management Strategies for the Classroom, 10*. http://www.acel.org.au/acel/ACEL_docs/Publications/e-Teaching/2016/e-Teaching_2016_10.pdf. Accessed December 2018.

Pink, D. H. (2010). *Drive: The surprising truth about what motivates us*. Edinburgh: Canongate Books. ISBN 10-1786891700.

Pugh, S. (2014). Getting down to business. *Education in Chemistry, 51*(5), 18–21. http://www.rsc.org/eic/2014/09/business-awareness-undergraduate-teaching. Accessed December 2018.

Pugh, S. (2017). Teaching career skills to undergraduates. *Education in Chemistry, 54*(1) https://eic.rsc.org/feature/teaching-career-skills-to-undergraduates/2500262.article. Accessed December 2018.

QAA. (2014). Subject benchmark statement: Chemistry. https://www.qaa.ac.uk/docs/qaa/subject-benchmark-statements/sbs-chemistry-14.pdf?sfvrsn=99e1f781_14. Accessed December 2018.

Riley-Douchet, C. (1997). A three-step method of self-reflection using reflective journal writing. *Journal of Advanced Nursing, 25*(5), 964–968.

Royal Society of Chemistry (RSC). (2017). Accreditation of degree programmes. http://www.rsc.org/images/Accreditation%20criteria%202017-%20update%20july%2017_tcm18-151306.pdf. Accessed December 2018.

Royal Society of Chemistry (RSC). (n.d.). Tackling global challenges. www.rsc.org/campaigning-outreach/global-challenges/. Accessed September 2018.

Rule, A. (2006). Editorial: The components of authentic learning. *Journal of Authentic Learning, 3*(1), 1–10.

Seery, M. (2015). Putting chemistry in context, *Education in Chemistry, 52*. https://eic.rsc.org/feature/putting-chemistry-in-context/2000106.article. Accessed December 2018.

Stice, J. E. (1987). Using Kolb's learning cycle to improve student learning. *Engineering Education, 77*(5), 291–296.

Tomlinson, M. (2012). Graduate employability: A review of conceptual and empirical themes. *Higher Education Policy, 25*, 407–431.

Wilkinson, D., & Aspinall, S. (2007). *An exploration of the term 'Commercial awareness': What it means to employers and students*. Coventry: National Council for Graduate Entrepreneurship.

Chapter 12
RAw Communications and Engagement (RACE): Teaching Science Communication Through Modular Design

Martin McHugh, Sarah Hayes, Aimee Stapleton and Felix M. Ho

Abstract This project is motivated by an increasing demand from public bodies and research funding agencies for outreach and public engagement, believing this to be a path towards enhanced public understanding, recruitment and research with impact. Yet many STEM (Science, Technology, Engineering and Mathematics) graduates and professionals lack the appropriate communication skills required to engage with the public. To address some of these issues, the project RACE (RAw Communication and Engagement) was jointly initiated by universities and industrial partners across Europe. Through the design and implementation of adaptable modules incorporating content knowledge, scientific communication and public engagement skills, this international project aims to equip students and researchers alike to ground their work within the wider global society and communicate their research with public audiences. A key feature is the direct incorporation of actual public engagement activities into RACE modules, for the mutual benefit of participants and wider society. Given this, the chapter will take a practitioner perspective to shed light on the inner workings of a modular scientific communication course. Core issues of engagement, bridging theory and practice, evaluation and collaboration are all highlighted through reflections and research data from implementing the RACE programme as a Ph.D. summer school with doctoral-level students. The chapter is concluded by bringing forward core teaching and learning protocols that are integral to running the RACE programme.

M. McHugh · S. Hayes · A. Stapleton
Synthesis and Solid State Pharmaceutical Centre, University of Limerick, Limerick, Ireland
e-mail: Martin.McHugh@ul.ie

S. Hayes
e-mail: Sarah.Hayes@ul.ie

A. Stapleton
e-mail: Aimee.Stapleton@ul.ie

F. M. Ho (✉)
Department of Chemistry - Ångström Laboaratory, Uppsala University, Uppsala, Sweden
e-mail: Felix.Ho@kemi.uu.se

M. Schultz et al. (eds.), *Research and Practice in Chemistry Education*,
https://doi.org/10.1007/978-981-13-6998-8_12

Introduction

The rationale for the pursuit of scientific endeavours, specifically within STEM education, is diverse (Osborne, Simon, & Collins, 2003). There is the notion of responsibility on the part of scientists to contribute to economic performance and grand societal challenges (e.g. Gago et al., 2004; UNCTD, 2017). STEM increasingly shapes our lives. With this in mind, there is a growing suite of scientific theories and ethical arguments that can only be dealt with through a public understanding of science. People require a higher level of scientific literacy than ever before to live and run their daily lives (Bromme, Thomm, & Wolf, 2015). Against this backdrop, demands are placed on scientists to act on their responsibility to communicate research impacts in a two-way public dialogue (Duschl, Schweingruber, & Shouse, 2007). This is grounded in the idea of developing the socio-scientific citizen, one who can interact with science through debate and everyday decision-making (Dillon, 2009). Policy has also picked up on this, with funding bodies and research institutes increasingly requiring education and public engagement (EPE) to be a vehicle towards delivering research with impact (Selin et al., 2017; Van Oudheusden, 2014).

The above falls in line with Responsible Research and Innovation (RRI), which refers to the ethical need for science to impact wider society in a continuous and beneficial manner. Efforts to ensure transparency of research are abound. However, transparency represents an uphill battle with Sutcliffe (2011, p. 6), noting that public trust in science is 'lost', potentially due to research and innovation often being uncertain with unknown consequences (Owen, Macnaghten, & Stilgoe, 2012).

The current landscape requiring scientific research practice to support and facilitate public engagement (Entradas & Bauer, 2017) has resulted in new interest in communication and outreach. While calls for such scientifically based communication with the wider society are ongoing, studies suggest that scientists in general lack the relevant skills for effective interaction with the public, and education and training opportunities are not widely available (Woods-Townsend et al., 2016; Edmondston & Dawson, 2014). Such education and training opportunities are vital, as meaningful communication with the public is otherwise dependent on the innate ability of the researcher or scientist (Bray, France, & Gilbert, 2012; Bauer & Jensen, 2011).

Addressing such needs is the core motivation for the project RAw Communication and Engagement (RACE). Initiated in 2015 and funded by the European Institute of Innovation & Technology (EIT) Raw Materials, RACE is a joint international project between six university and two industrial partners across Europe. The goal of RACE is to address communication and public engagement requirements among university students and working professionals in STEM disciplines related to raw materials. Rather than restricting the modules to purely educational settings, the RACE programme incorporates actual outreach and public engagement activities as value-added components for the mutual benefit of real societal stakeholders and course participants.

This chapter describes the design and implementation of the education and outreach initiatives developed during the RACE project. We aim to strike a balance between theoretical/research perspectives and more implementation-oriented discussions that would be valuable for practitioners. This highlights a range of issues to consider when developing modules aimed at improving participants' skills in scientific communication and public engagement.

RACE Programme Design

RACE is a cross-national and cross-sectoral project with the following partners: University of Limerick (UL), Uppsala University (UU), University of Eastern Finland (UEF), the Technical University of Madrid (UPM), Boliden Tara Mines Ltd. and Rusal Aughinish Alumina Ltd. With this, its design is a collaborative effort between multiple experts across a multitude of disciplines. This process is described presently.

Baram-Tsabari and Lewenstein (2017) describe a proliferation of science communication courses, ranging from one-off lectures, day and week-long events to core modules in university courses. However, definitions, content and context pertaining to science communication education and training are highly diverse. As such, creating a science communication course based on the current literature is a difficult task. Given this remit, the RACE programme design involved a workshopping process in which the curriculum developers looked externally, but also relied on their experience to create a course. A key consideration from the outset was the need for a course that is of high quality, modular and adaptable.

The first step involved the design of a module that is transferable to multiple contexts, which was of central importance in order to be able to accommodate the range of backgrounds among academic and industry partners. Moreover, the training of higher education staff to deliver the module was also a goal from outset. This model is summarised in Fig. 12.1.

Two key areas were initially identified as pillars for the module design. The first pillar was education in the technical skills relevant to STEM communication.

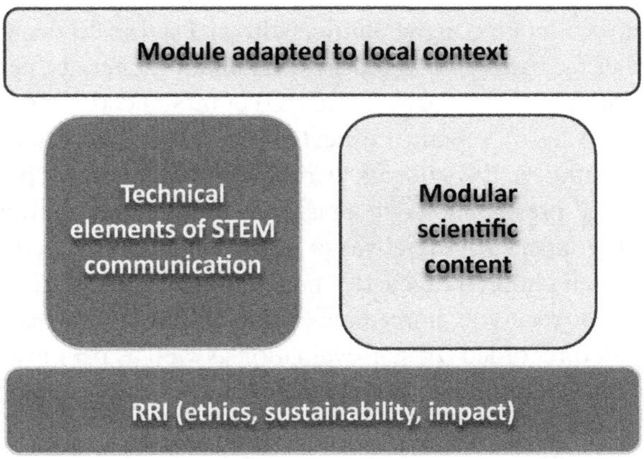

Fig. 12.1 Model informing the design of RACE modules with core (dark grey) and adaptable (light grey) content

Development of content for this pillar was informed by research, indicating that there are issues with how researchers communicate, with this being very much dependent on their abilities to present, explain and engage. The second pillar was modular scientific content. The scientific content pillar is adaptable to the local educational context so that the professional educational or training outcomes are met. This was necessitated by the diverse backgrounds of the project partners as well as the range of requirements associated with educational programmes at the local level. In the case of RACE, the topic of 'raw materials' was the unifying theme of this content pillar but is a theme that could readily be exchanged for another topic area. Underpinning all of this were considerations of ethics, sustainability and impact relevant to RRI. It was important to establish an explicit relationship with the concept RRI, which describes an inclusive approach to the process of research and innovation with communication between wide-ranging stakeholders throughout. This was guided by work from the European Commission, which has created a normative framework for RRI involving six policy agenda areas: ethics, open access, public engagement, gender equality, governance and science education (European Commission, 2014).

To fully flesh out the content of the course, the designers took part in a brainstorming process over two days. Necessary strands that emerged for participant learning were reflection, real-world assessment and peer feedback. In line with this, the philosophy underpinning RACE was a cycle of:

- Interactive learning
- Guided and independent assessment
- Self-reflection and evaluation.

To test this curriculum content and framework, a pilot of the module was developed as a part of the Masters of Science in Chemical Engineering course 'Materials for Sustainable Development' at UU. The implementation and evaluation of this Masters module has been described in Hayes, Brandell, and Ho (2018), and a summary is provided below.

The implemented module consisted of three seminars. The first considered raw materials, their range of use and applications. The second included considerations of RRI more explicitly, and students were asked to conduct a debate on topics that involve competing scientific, industrial and socio-economic interests (e.g. state subsidies for battery-powered cars, tax on raw materials, new mines for rare earth metals). The final component introduced communication and engagement. After a seminar focusing on practical aspects of scientific communication and public engagement techniques, the students were given an assignment involving three-minute 'elevator pitch' presentations on an assigned topic with relevance to raw materials. Working in groups, each topic was to be presented to three different categories of audiences: school children (aged 15), potential investors and a research funding body. In addition to receiving immediate feedback from course instructors, the presentations were video-recorded for self-evaluation as well as peer evaluation by others in the group. This aided students both in their learning and in creating a reflective piece of writing on the experience.

Table 12.1 Paired sample test comparing pre- and post-course mean values (5-point Likert scale 2017 cohort, $N = 43$, $df = 42$; Cronbach's alpha α calculated on post-course responses; n_i = number of items grouped in the theme)

Theme	Pre	Post	Paired difference mean	Standard error	p	Effect size r
Understanding of raw material ($\alpha = 0.72$; $n_i = 12$)	3.64	3.92	0.27	0.051	<0.001	0.63
Scientific communication ($\alpha = 0.62$; $n_i = 12$)	3.50	3.62	0.13	0.060	0.041	0.31
Individual responsibility ($\alpha = 0.76$; $n_i = 4$)	3.40	3.61	0.22	0.093	0.025	0.34
Ethics and governance ($\alpha = 0.63$; $n_i = 8$)	3.51	3.75	0.24	0.060	<0.001	0.52

Evaluation using a pre-/post-questionnaire, the adapted form of an instrument utilised by the FP7 project *Irresistible* (Blonder, Rap, Zemler, & Rosenfeld, 2017) showed that the module together with the Master's course had statistically significant positive effects on the students' self-reported understanding of raw materials, scientific communication skills, views on individual responsibility, as well as views on issues of ethics and governance (Table 12.1). As the module was incorporated into an existing course [2 out of 10 credit (ECTS) points], these effects could not be isolated to the module intervention alone, but the positive outcomes from the Masters course nevertheless supported the continued use of the overall module design concept. A systematic reflection of the aspects of the course that were effective and parts that required modification provided key findings that formed the basis for the design and implementation of an expanded module at the Ph.D. level. This work is described in the sections below.

Setting the Scene: Ph.D. Summer School and Research Context

The core intervention from the RACE programme was the Ph.D. module (6 ECTS credits of the European Credit Transfer System) that was implemented jointly by each of the university partners of the RACE project, the structure of which is illus-

Fig. 12.2 The structure of the RACE Ph.D. module (6 ECTS)

trated in Fig. 12.2. A key differentiating feature of the Ph.D. module over the one implemented in the Master's course is the expansion of the core pillar of science communication into a joint week-long residential summer school, where the RRI underpinnings were also incorporated. The pillar of raw material content, as well as formal assessments of the participating Ph.D. students, was instead covered separately by the individual university partners through reading assignments and seminar series prior to and after the residential summer school. This allowed for accommodation of the details of subject content and assessment in accordance with the local educational requirements of each institution, in line with the module design model discussed above (Fig. 12.1). Moreover, the assessment tasks included real public engagement events conducted within local communities associated with each university partner (see sections below).

The Residential Summer School—The Scientific Communications Common Core

The Ph.D. residential summer school took place at UL in August 2017 with participation from 24 Ph.D. students from UL (10), UU (10) and UEF (4). The timetable for the week is displayed in Table 12.2.

The summer school had the aim of enabling participants to design, organise, deliver and evaluate a range of public engagement activities with various audiences. To enable this, the crux of the delivery model chosen, consisted of interactive, experiential and active learning sessions, presented with supportive opportunities to interact, learn and reflect. Ten core teaching and learning areas (Table 12.3) identified during the design of the programme laid the foundation and central thinking of the summer school.

In addition to considering these core areas, the implementation of the summer school took advantage of the resources and strengths of the host and participating institutions. For example, one of the teaching sessions looked at variety in presenting. Within this session, multiple presentation types were modelled by guest speakers including a Thesis-in-Three presentation that is popular in the Irish context. Public engagement can take many forms, and the merging of art and science is always valuable. This was explored through the public screening of the film 'The Farthest'

Table 12.2 Ph.D. summer school timetable

Time	Monday (Day 1)	Tuesday (Day 2)	Wednesday (Day 3)	Thursday (Day 4)	Friday (Day 5)
9:30–10:30	Why communicate and engage	How to develop creative direct EPE events	Circus of demonstrations	Engaged research and where it can lead How to be engaged in your own research	Group works and discussion: putting actions into practice
10:30–11:00	Coffee break	Coffee break	Coffee break	Coffee break	Coffee break
11:00–13:15	Introduction to communication and engagement	Presenting demonstrations	Trying your own demonstration	Types of presentations Thesis-in-Three Funding agency School audience	Group works and discussion: putting actions into practice
13:15–14:15	Lunch	Lunch	Lunch	Lunch	Lunch
14:15–15:45	How to be engaging 1	Presenting workshops	Trying your own demonstration	OPCW: presenting to policy-makers	
15:45–16:00	Coffee break	Coffee break	Coffee break	Coffee break	Coffee break
16:00–16:45	How to be engaging 2	Organising direct EPE events	Writing and presenting for different audiences	Entrepreneurship and elevator pitches	
16:45–17:30	How to be relevant		Writing articles and blogs		
20:00	'The Farthest' screening		Traditional Irish Music	Group meal	

(an Irish Emmy award winning documentary about NASA's voyager programme) that coincided with the summer school, including a question and answer session with the director, for which the RACE course coordinator had been invited as the guest moderator. This highlighted how collaboration can bring the message and dialogue to a much larger audience. This variety was strengthened by the addition of a video conference call between the RACE participants and a senior scientific advisor at the Organisation for the Prohibition of Chemical Weapons (OPCW) on how they engage with policy-makers around science, with the scientific advisor having been a colleague of one of the summer school instructors. Moreover, when participants were asked to observe and test out their own demonstrations, this was done in a bespoke manner depending on the speakers and equipment that were available locally at UL.

Table 12.3 Core areas of the RACE summer school curriculum along with a description of each area

Core area	Description	Summer school sessions
Motivation and challenges in public engagement	Introduction to the course providing a rationale for public engagement. Explicit discussion about what the barriers are, and why it is worth pushing past them to develop public dialogue. The role of Education and Public Engagement (EPE) in research along with the difficulties in performing EPE activities including ambassador and audience attitudes, diverse contexts, the public deficit model, the appropriate use of scientific language and conceptual understanding among audience members (Bauer & Suerdem, 2016; Markic & Childs, 2016; Bauer & Jensen, 2011; Johnstone, 2000; Cleaves, 2005; Osborne et al., 2003)	Day 1: Why communicate and engage, Introduction to communication and engagement
Responsible Research and Innovation (RRI)	Extensive examination of RRI and the researchers' ethical responsibility to communicate to the public, particularly when the public often act as a funding body. Exemplifying the benefits of communication and relevant education and training	Day 1: Introduction to communication and engagement Day 4: Engaged research and where it can lead
Engagement techniques and methods	How a presenter can grab the attention and interest of an audience, underscoring the utilisation of hooks (e.g. demonstrations, discrepancy, conceptual change, active learning, questioning and implementing novel technologies; McCrory, 2011; McHugh & McCauley, 2015)	Day 1: How to be engaging 1, How to be engaging 2
Relating to your audience	Outlining why presentations are important to an audience. Presenters need to create a quick connection and establish relevance to the audience's daily lives. Audiences prefer presenters with whom they can relate (Bultitude, 2011; Besley, Dudo, Yuan, & Abi Ghannam, 2016)	Day 1: How to be relevant Day 3: Writing and presenting for different audiences
Explaining your ideas	How to concisely present ideas without confusing an audience. Accuracy and simplicity are outlined with reference to scientific versus everyday language. Examples of analogies are used but only if they accurately draw a true comparison with the science and not create misconceptions. In addition, the importance of considering the background and interests of the audience is addressed both at a general level, but also exemplified in specific role-playing exercises and seminars, covering a diverse range of potential audiences, from primary age school children, to teenagers and adults of the general public, to potential venture capitalists and diplomats	Day 1: How to be engaging 1 Day 2: Presenting demonstrations, Presenting workshops Day 4: Entrepreneurship and elevator pitches

(continued)

Table 12.3 (continued)

Core area	Description	Summer school sessions
Developing public engagement activities	Identifying the purpose of public engagement activities and the potential constraints associated with them. Boundaries need to be established so that interactions are appropriately utilised to communicate with clarity. Idea refinement is explored along with the importance of practice and rehearsals	Day 2: How to develop creative direct EPE events
Planning and logistics	Consideration of the purpose of EPE events from a presenter's perspective, venues, slides, projectors, stage, etc. Moreover, thinking about the audience, their logistics, child protection concerns, health and safety, parking and transport	Day 2: Organising direct EPE events
Conducting demonstrations	Providing insight into the importance of demonstrations and practice, including potential pitfalls and problems and how to deal with them should they arise. Engaging the audience as active participants and stakeholders in the progress of any demonstration through delivery methods such as prediction–observe–explain (POE; White & Gunstone, 1992; Crouch, Fagen, Callan, & Mazur, 2004)	Day 2: Presenting demonstrations Day 3: Circus of demonstrations, Trying your own demonstration
Conducting workshops	The process of guiding an audience to facilitate enquiry-based learning. Explaining how to give concise instructions in a systematic manner so that an audience will understand. Moreover, participants developed and practiced a simple demonstration to convey an aspect of their research	Day 2: Presenting workshops Day 3: Trying your own demonstration
Evaluation of public engagement	An examination of evaluation techniques and their appropriate implementation. Defining the purpose of evaluation and choice of appropriate evaluation methods accordingly. Data interpretation and cyclical improvement in EPE events are explored (Jensen, 2014)	Day 4: How to be engaged in your own research Day 5: Group works and discussion: putting actions into practice

The Modular Content and Context—Raw Materials

As illustrated in Fig. 12.1, key features of the course design framework were the ability to adapt to a particular context and to interchange scientific topics/themes. For the purposes of this particular programme, the modular context was that of raw materials, with the focus varying across partner institutions (e.g. raw materials for pharmaceuticals in UEF and UL, renewable energy and manufacturing processes in UL and UU, advanced materials and energy storage at UU, raw materials in the context of mining in UPM). Students accepted into the Ph.D. module were given reading material and took part in seminars prior to the residential summer school at their local institution. By having a unifying content theme of raw materials for all participating institutions, but not with identical and prescribed material, it was possible to tailor the content to the local institutional and education requirements of the participants. Clearly, in other implementations of this framework, the topic of raw materials can be changed to suit local educational needs while keeping the core pillar of scientific communication and public engagement underpinned by the principles of RRI.

Another aspect of modularity was regarding the context of implementation. An explicit aim was to create a programme that could support a diverse range of institutions, reflective of the RACE project teams. To achieve this, consultation with potential trainees was vital during the planning stages to set out their needs. In applying for this course, participants were in a sense self-selecting. As such, they had an understanding of the benefits of the course (including potential academic recognition), their own goals and aspirations for new knowledge. Additionally, the module was designed to be adaptive, with multiple training models possible for the pre- and post-summer school components (e.g. lecture/seminar series, training days, online training and week-long courses). This was also true of the assessment and examination, which was decided upon and carried out by the individual institutions. With any assessment, authentic contexts should ideally be integrated, and indeed multiple EPE events were conducted at each of the institutions as part of assessment. This adds to the bespoke nature of the course as every EPE context is vastly different. At present, a total of 24 EPE events have been staged across four countries as part of the Ph.D. summer school assessment. Examples will be explored in more detail in the emergent results section below.

Evaluation of the Summer School

Research Methodology

Following a successful research ethics application at UL, the Ph.D. module ($n = 24$) was evaluated using a research methodology employing exit cards (utilising post-its), semi-structured qualitative interviews, observations and reflections among the course delivery team. The exit cards provided for real-time qualitative data that allowed for question development in the semi-structured interview tool and points of interest for reflection. The interviews and reflections enabled a more in-depth insight after the intervention and subsequent module assessment took place. This qualitative research scheme was employed to find out in-depth viewpoints from participants. Multiple methods were employed in an effort to triangulate data while each participant was treated as a singular case in a multiple case study framework. Yin (2014) advocates this approach in the creation of replication logic that can produce robust insights into lived experiences. The exit card strategy utilised post-it walls during the residential summer school to give the programme designers ‘real time’ information, so the adaptations could potentially be instigated 'on the fly' if needed. This represented a simple and efficient way to collect evidence (Barker, 2002) at a singular moment in time (Thomas, 1999). A total of nineteen semi-structured interviews were conducted after the Ph.D. module was completed. This approach is seen as being beneficial as it allows participants to share their experience in an individual manner while also having a structure to aid cross-case comparison (Wahyuni, 2012; Cohen, Manion, & Morrison, 2011). The interview schedule has 15 questions; however, in line with the iterative nature of qualitative enquiry, questions can be amended, added or removed to foster understanding and insight.

Data analysis is ongoing, utilising a coding framework set out by Saldaña (2012). Key steps in the process include data familiarisation, translation, writing analytical memos, first and second cycle coding. First cycle coding employs elemental coding with descriptive, in vivo, and process or gerund codes. The aim is to fragment the data and gain a key insight into trends and patterns. After this, thematic analysis will be used during second cycle coding to amalgamate and refine the codes so that a valid and robust interpretation of the data can be achieved (Clarke & Braun, 2013). In the following section, we present emergent results from the data analysed so far comprising transcription and first cycle coding of all exit cards and approximately half of the interviews.

Emergent Results

Emergent results from the study indicate that participants were unaware of the amount of work involved in the development of an EPE activity. Pelger and Nilsson (2018) note that a successful grounding to science communication is built on practice. It seems as people who mostly have been on the receiving end of presentations and lectures all their lives, they had never been given a true insight into the work behind the scenes. This came to the forefront of their minds when they conducted their own EPE events in which young audiences pushed the boundaries of 'the presentation plan'.

> …we practiced the day before and showed each other, but when the moment came for the pupils, it was like totally different story! *Amy*

> I realise how much preparation goes into these talks, before I sort of thought, well I knew people are prepared, but I didn't know that people were this prepared. *Christy*

Along with the above recognition was the core role of 'playing' to the audience. Participants identified the need to know the people in front of you and plan around their knowledge, needs and ability. Again, this is evidenced with reference to the interviews.

> Assessing the target audience, that's the first thing I think about, I never thought about this, but now it's like, ok, who am I talking to, can they understand? *Dave*

> I never usually spend time thinking about how important it can be communicating with a general audience but also with an audience outside academia, that, I never really considered before… *Frank*

Some would consider the audience as the predominant factor in science communication efforts (Bray et al., 2012). The audience is often referred to as something with fleeting attention that needs to be 'captured'. Building a relationship with the audience throughout EPE is vital so that expert science can be understood by the varying levels of knowledge present in an audience (Bray et al., 2012). Moreover, the evidence above moves away from the public deficit model of science communication. The participants recognised the audience and their role in a two-way scientific dialogue (Besley et al., 2016). This was a large component of the Ph.D. summer school and demonstrates the practical importance of the programme as a whole. Acknowledging the needs of the audience is something that was mentioned in 55% of the exit cards which asked participants about their learning from the course.

Pertaining to further learning, skill development was also a key factor noted throughout interviews and exit cards. Skills such as the use of hooks, adapting language and simplifying concepts were referenced the most. Students seemed to realise this through a process of reflection.

> If I was to do a demo without the course, I wouldn't have really known how to guide it or really direct it, like I wouldn't have thought of using analogies to explain polymers. *John*

> I have not really given thought of how the details of communicating science while I want to communicate science, perhaps I lacked some tools, I am not sure. I have the skills to do this after the course. *Amy*

In the light of this, many of the Ph.D. students noted that they lacked particular tools to create engagement and that the RACE programme provided them with this skill set. The Ph.D. summer school gave students a new engagement 'toolkit', recognised by Bultitude (2011) as a core aspect of science communication. The author draws together the most important areas of science communication including acknowledging the audience, evaluating oneself, learning from others, practice and thinking creatively. These are important aspects that are covered in the RACE module. Importantly, there was a recognition of skill development among participants especially when they enacted their own EPE events and fell back upon their learning.

The results so far illustrate the transformative potential of active learning in the public engagement realm with a total of 24 real-world EPE events conducted across UU, UL, UPM and UEF, some of which constituted part of the participants' examination for the Ph.D. module. The contexts provided for diversity in the types of events are developed and implemented. For example, UU had multiple school collaborations and in-house chemistry workshops in laboratories. UU also put a twist on two events that currently take place in Ireland. They modified the established 'Pint of Science', an event where science is presented in a pub or bar environment, to 'Science and Beer'. Additionally, they also held a 'Thesis-in-Three' event for undergraduate and Master's students at the university, inspired by the multiple presentation formats held during the Ph.D. summer school. UPM conducted multiple workshops on a variety of topics, from drones, to mining and critical raw materials. Moreover, the audiences varied heavily in these activities from children to adults. They also took part in 'Expominerales', a public event held outside where people have the opportunity to pan for gold, use metal detectors and explore types of minerals. UEF conducted an open evening for the public about pharmaceuticals. They took part in a national event, 'SciFest', and directed a workshop. They also conducted their own event at a shopping centre with posters and interactive stands. In addition to this, UEF published an article in a professional pharmaceutical journal that gave an overview of the drug discovery process in accessible terms. Finally, UL took part in university open day demonstrations. They also took part in 'Physics Busking' in a local food market, where physics experiments are performed in front of and with the public. School visits and workshop were staged, while science shows were also performed during the Irish 'Science Week'. Engaging in EPE is effort and resource intensive, but is also a personal learning journey that put the participants' knowledge and skills to the test. In the results, much of the important learning noted by participants occurred through the execution of real-world EPE activities. The majority of participants relied upon what they had learned in the RACE programme, and this demonstrates the practical value of the teaching. Furthermore, this also bolsters the implementation of interviews after the assessment as the exit cards, although useful, could not bring this learning to the fore. In summary, the RACE creators and participants went through a similar process of learning, implementation and reflection to achieve largest knowledge gains. This will be explored more thoroughly in the next section with reference to the RACE project as a whole.

Practitioner Reflections and Implications for Practice

The following section will describe the practitioner learning and development that took place over the iterative progression of the RACE programme. The intention is that this reflective process will facilitate the implementation and adaptation of similar modules built upon the RACE module framework. By design, the RACE framework supports transferability to contexts through modularity while keeping science communication and RRI at its core.

After the Ph.D. summer school took place, the RACE coordinators conducted a group reflection to summarise learning and modifications for future iterations of the course. Pertaining to the coordinators' performance and the lectures delivered, a key feature that would prove beneficial is a summary aspect. Students seemed to need their learning summarised so that they could reflect upon key points and bring them forward. Throughout presentations, modelling of good science communication practice was abound, but signposting to clearly articulate and reinforce specific principles and practices throughout the training is recommended to sustain learning. Both the Ph.D. and Masters students had a highly scientific background, but science communication does not represent their comfort zone. Particularly, it is a new subject area, and they struggled at times with the content. Given this remit, we would suggest that pedagogical scaffolding is required to ease students into the course. A core positive is the recommended cycle of learning, implementation and reflection. Participant learning from this method was observed on multiple occasions and referenced throughout results thus far. This proves to be a useful and transferable result, one that is key to delivering the programme. Complementing this is the need for reflection among participants to complete the proposed learning cycle underpinning the central aspect to the RACE programme. According to Gibbs (1988, p. 9), 'It is not sufficient to have an experience in order to learn. Without reflecting on this experience, it may quickly be forgotten, or its learning potentially lost'. Reflection came to the fore in the student interviews where they could easily recall their learning and its application in a variety of settings, not just science communication. Given this, we believe the content to be beneficial, practical and relevant, but an awareness that participants in science communication education and training courses are generally novices is warranted.

With this background, we would advise dividing laboratory or more practical sessions over multiple days so that there is a mix between theory and practice. Having diverse content was something that was thought about before the running of the school as a number of guest speakers were brought in to bring a new voice to the content. A teaching session where this was paramount involved role-playing. Participants were asked to present a scientific demonstration, while the audience was asked to act like a particular type of audience, i.e. young children or students. This gave participants valuable insights and a realisation into some of the problems and opportunities an audience may present. In alignment, we would promote such active pedagogies that are built around the communicator and audience dynamic. This also acted as stepping stone, a sort of safe environment in which to make mistakes and learn. This is highly relevant as the 'handholding' of participants that took place at the start of the week slowly released. Of course, the crux of the model that proved most beneficial is that students would conduct their own EPE event as part of their assessment.

As can be seen from the above, a diversity of EPE events were put in place. This is beneficial for multiple reasons. Firstly, participants can adapt their particular skill set to a communication format that suits them. For example, writing an article for a lay audience is very different to planning and running a workshop. Furthermore, many events are conducted in teams, so all of the responsibility is not just placed on one person. Secondly, the array of events illuminates the modular aspect of the course. University partners often utilised established local opportunities to conduct events, for example, 'Science Week' in Ireland or 'SciFest' in Finland. Thirdly, the topic content is different to suit the background expertise of the researchers and relevant institutions. UEF conducted a lot of work with pharmaceuticals and UPM with minerals and mining. They both align under the raw materials umbrella, but also reflect their specific interests and more importantly capabilities. Indeed, the interviews reveal the importance of learning in real-world environments. The assessment holds a broad relevance and in our view is fundamental to any science communication course. It caps off the learning cycle the RACE programme promotes along with the modular and transferable nature of all our science communication learning goals.

In summary, the course and its implementation were positive and at this stage small modifications are recommended. As a whole, RACE as it stands provides for a well-developed and tested science communication course. Interestingly, the project team followed the same learning progression as the students: knowledge acquisition, implementation and reflection, with reflection being a fundamental component. Teaching the idea of reflection to scientific research students can be difficult; however, the assessment whereby participants have to plan and implement their own EPE event enabled the students to firstly reflect and secondly see the benefit of such practice. As educationalists, we are very aware that 'assessment is the tail wagging the dog'. The Ph.D. summer school assessment was truly grounded in the content of the course, and this is a key positive for any similar intervention.

Conclusions and Implications for Practice

This chapter describes the implementation of the RACE programme, with insights into data. The narrative demonstrates an increased demand from multiple standpoints for effective science communication. The implementation of RACE has a multitude of benefits to participants' interpersonal and transferable skills. Developing this skill set is proved to be more complex than participant's initial perceptions. Assessment in an authentic setting brought this to light, along with a pedagogical model, which aimed to bridge the theory practice divide through a fluid learning, implementing and reflecting loop. RACE will be implemented again to continue the programme refinement and will strive to connect with alumni to ascertain if the benefits of the course are upheld on a longitudinal basis. In terms of longevity, the teaching and learning material from the project is archived on a website at https://eitrawmaterials-race.com/ with the goal of facilitating the dissemination of the learning throughout RACE.

Acknowledgements This project is funded by the EIT Raw Materials KIC Project 15028, No. FPA 2016/EIT/EIT Raw Materials. This publication has also received support through the Synthesis and Solid State Pharmaceutical Centre, which is funded by Science Foundation Ireland (SFI) and co-funded under the European Regional Development Fund under Grant 12/RC/2275, as well as the Centre for Discipline-Based Education Research in Mathematics, Engineering, Science and Technology (MINT) at Uppsala University. We would also like to thank the significant contribution of Prof. Daniel Brandell, Uppsala University, and Dr. Paul McCrory of *Learn Differently Ltd.* in the implementation of the project.

References

Baram-Tsabari, A., & Lewenstein, B. V. (2017). Science communication training: What are we trying to teach? *International Journal of Science Education, Part B, 7*(3), 285–300. https://doi.org/10.1080/21548455.2017.1303756.

Barker, N. C. (2002). Foundations in qualitative methods. In C. Barker, N. Pistrang, & R. Elliott (Eds.), *Research methods in clinical psychology: An introduction for students and practitioners* (pp. 5–72). Chichester, UK: Wiley London.

Bauer, M. W., & Jensen, P. (2011). The mobilization of scientists for public engagement. *Public Understanding of Science, 20*(1), 3–11. https://doi.org/10.1177/0963662510394457.

Bauer, M. W., & Suerdem, A. (2016). Developing science culture indicators through text mining and online media monitoring. In *OECD Blue Sky Forum on Science and Innovation Indicators 2016*, 19–21 September 2016. Ghent, Belgium: OECD.

Besley, J. C., Dudo, A. D., Yuan, S., & Abi Ghannam, N. (2016). Qualitative interviews with science communication trainers about communication objectives and goals. *Science Communication, 38*(3), 356–381. https://doi.org/10.1177/1075547016645640.

Blonder, R., Rap, S., Zemler, E., &. Rosenfeld, S. (2017). Assessing attitudes about responsible research and innovation (RRI): The development and use of a questionnaire. *Sisyphus Journal of Education, 5*(3), 122–156.

Bray, B., France, B., & Gilbert, J. K. (2012). Identifying the essential elements of effective science communication: What do the experts say? *International Journal of Science Education, Part B, 2*(1), 23–41. https://doi.org/10.1080/21548455.2011.611627.

Bromme, R., Thomm, E., & Wolf, V. (2015). From understanding to deference: Laypersons' and medical students' views on conflicts within medicine. *International Journal of Science Education, Part B, 5*(1), 68–91. https://doi.org/10.1080/21548455.2013.849017.

Bultitude, K. (2011). The why and how of science communication. In P. Rosulek (Ed.), *Science communication* (pp. 31–58). Pilsen: European Commission https://www.scifode-foundation.org/attachments/article/38/Karen_Bultitude_-_Science_Communication_Why_and_How.pdf. Accessed December 2018.

Clarke, V., & Braun, V. (2013). Teaching thematic analysis: Overcoming challenges and developing strategies for effective learning. *The psychologist, 26*(2), 120–123. http://eprints.uwe.ac.uk/21155/.

Cleaves, A. (2005). The formation of science choices in secondary school. *International Journal of Science Education, 27*(4), 471–486. https://doi.org/10.1080/0950069042000323746.

Cohen, L., Manion, L., & Morrison, K. (2011). *Research methods in education*. London; New York: Routledge.

Crouch, C., Fagen, A. P., Callan, J. P., & Mazur, E. (2004). Classroom demonstrations: Learning tools or entertainment? *American Journal of Physics, 72*(6), 835–838. https://doi.org/10.1119/1.1707018.

Dillon, J. (2009). On scientific literacy and curriculum reform. *International Journal of Environmental and Science Education, 4*(3), 201–213.

Duschl, R. A., Schweingruber, H. A., & Shouse, A. W. (Eds.). (2007). *Taking science to school: Learning and teaching science in grades K-8*. Washington, D.C.: The National Academies Press. https://doi.org/10.17226/11625.

Edmondston, J., & Dawson, V. (2014). Perspectives of science communication training held by lecturers of biotechnology and science communication. *International Journal of Science Education, Part B, 4*(2), 195–210. https://doi.org/10.1080/21548455.2013.793433.

Entradas, M., & Bauer, M. M. (2017). Mobilisation for public engagement: Benchmarking the practices of research institutes. *Public Understanding of Science, 26*(7), 771–788. https://doi.org/10.1177/0963662516633834.

European Commission. (2014). *Responsible research and innovation: Europe's ability to respond to societal challenges*. https://ec.europa.eu/research/swafs/pdf/pub_rri/KI0214595ENC.pdf. Accessed November 25, 2018.

Gago, J. M., Ziman, J., Caro, P., Constantinou, C., Davies, G., Parchmann, I., et al., (2004). *Increasing human resources for science and technology in Europe*. Brussels: European Commission.

Gibbs, G. (1988). *Learning by doing: A guide to teaching and learning methods*. Oxford: Further Education Unit Oxford Brookes. https://thoughtsmostlyaboutlearning.files.wordpress.com/2015/12/learning-by-doing-graham-gibbs.pdf. Accessed December 2018.

Hayes, S., Brandell, D., & Ho, F. (2018). RAw communication and engagement. In O. E. Finlayson, E. McLoughlin, S. Erduran, S., & P. Childs (Eds.), *Electronic Proceedings of the ESERA 2017 Conference. Research, Practice and Collaboration in Science Education*, Part 17 (co-ed. Jenaro Guisasola), (pp. 2273–2282). Dublin, Ireland: Dublin City University. ISBN 978-1-873769-84-3.

Jensen, E. (2014). The problems with science communication evaluation. *Journal of Science Communication, 13*(1), 1–3. https://doi.org/10.22323/2.13010304.

Johnstone, A. H. (2000). Teaching of chemistry—Logical or psychological. *Chemistry Education Research and Practice in Europe, 1*(1), 9–15. https://doi.org/10.1039/a9rp90001b.

Markic, S., & Childs, P. E. (2016). Language and the teaching and learning of chemistry. *Chemistry Education Research and Practice, 17*(3), 434–438. https://doi.org/10.1039/c6rp90006b.

McCrory, P. (2011). Developing interest in science through emotional engagement. In W. Harlen (Ed.), *ASE guide to primary science education*. Hatfield: Association for Science Education.

McHugh, M., & McCauley, V. (2015). Designing physics video hooks for science students. *Physics Education, 51*(1), 015015. https://doi.org/10.1088/0031-9120/51/1/015015.

Osborne, J., Simon, S., & Collins, S. (2003). Attitudes towards science: A review of the literature and its implications. *International Journal of Science Education, 25*(9), 1049–1079. https://doi.org/10.1080/0950069032000032199.

Owen, R., Macnaghten, P., & Stilgoe, J. (2012). Responsible research and innovation: From science in society to science for society, with society. *Science and Public Policy, 39*(6), 751–760. https://doi.org/10.1093/scipol/scs093.

Pelger, S., & Nilsson, P. (2018). Observed learning outcomes of integrated communication training in science education: Skills and subject matter understanding. *International Journal of Science Education, Part B, 8*(2), 135–149. https://doi.org/10.1080/21548455.2017.1417653.

Saldaña, J. (2012). *The coding manual for qualitative researchers*. London: Sage Publications Ltd.

Selin, C., Rawlings, K. C., de Ridder-Vignone, K., Sadowski, J., Altamirano Allende, C., Gano, G., et al., (2017). Experiments in engagement: Designing public engagement with science and technology for capacity building. *Public Understanding of Science, 26*(6), 634–649. https://doi.org/10.1177/0963662515620970.

Sutcliffe, H. (2011). A report on responsible research and innovation. MATTER and the European Commission. https://ec.europa.eu/research/science-society/document_library/pdf_06/rri-report-hilary-sutcliffe_en.pdf. Accessed August 15, 2018.

Thomas, S. J. (1999). *Designing surveys that work! A step-by-step guide*. Thousand Oaks, CA: Corwin Press.

United Nations Conference on Trade and Development (UNCTD). (2017). *The role of science, technology and innovation in ensuring food security by 2030*. New York and Geneva: United Nations. https://unctad.org/en/PublicationsLibrary/dtlstict2017d5_en.pdf. Accessed December 9, 2018.

Van Oudheusden, M. (2014). Where are the politics in responsible innovation? European governance, technology assessments, and beyond. *Journal of Responsible Innovation, 1*(1), 67–86. https://doi.org/10.1080/23299460.2014.882097.

Wahyuni, D. (2012). The research design maze: Understanding paradigms, cases, methods and methodologies. *Journal of Applied Management Accounting Research, 10*(1), 69–80.

White, R., & Gunstone, R. (1992). *Probing understanding*. New York, NY: Routledge.

Woods-Townsend, K., Christodoulou, A., Rietdijk, W., Byrne, J., Griffiths, J. B., & Grace, M. M. (2016). Meet the scientist: The value of short interactions between scientists and students. *International Journal of Science Education, Part B, 6*(1), 89–113. https://doi.org/10.1080/21548455.2015.1016134.

Yin, R. K. (2014). *Case study research: Design and methods*. London: Sage Publications.

Chapter 13
Constructive Alignment Beyond Content: Assessing Professional Skills in Student Group Interactions and Written Work

Renée Cole, Gil Reynders, Suzanne Ruder, Courtney Stanford and Juliette Lantz

Abstract Many recent reports have called for the development of professional skills as part of a quality undergraduate experience to better prepare college graduates to enter the workforce; this preparation requires more explicit assessment of these skills in chemistry and other STEM fields. While professional skills may be part of programmatic goals, the development and assessment of these skills must largely take place as part of course experiences. However, skill development remains an implicit component of coursework that is not addressed in a meaningful way due to a lack of tools and strategies for assessing skill development and providing feedback to students. To address this need, rubrics were developed as part of the Enhancing Learning by Improving Process Skills in STEM (ELIPSS) Project to assess evidence of information processing, critical thinking, problem solving, communication, teamwork, management, and metacognition skills in student written work and group interactions. Here, we report a broad overview of how the rubrics have been implemented in multiple learning environments including detailed descriptions of two case studies. Results from faculty surveys and analysis of these case studies indicate that instructors have changed their curricula to include more explicit assessment of professional skills using a variety of instructional practices. Interviews with teaching assistants and students indicate that while the teaching assistants obtained a clearer understand-

R. Cole (✉) · G. Reynders
University of Iowa, Iowa City, USA
e-mail: renee-cole@uiowa.edu

G. Reynders
e-mail: gilbert-reynders@uiowa.edu

S. Ruder
Virginia Commonwealth University, Richmond, USA
e-mail: sruder@vcu.edu

C. Stanford
Ball State University, Muncie, USA
e-mail: clstanford@bsu.edu

J. Lantz
Drew University, Madison, USA
e-mail: jlantz@drew.edu

M. Schultz et al. (eds.), *Research and Practice in Chemistry Education*,
https://doi.org/10.1007/978-981-13-6998-8_13

ing of how to assess professional skills in both written work and group interactions, the students continue to need more specific feedback in order to improve their skills.

Introduction

Professional Skills

The development of professional skills as part of a quality undergraduate experience to better prepare college graduates to enter the workforce has been increasingly emphasized. Professional skills (also commonly referred to as soft skills, process skills, twenty-first century skills, transferable skills, or practical skills) such as information processing, critical thinking, problem solving, interpersonal communication, and teamwork have been identified as crucial in the hiring process for new employees, but these skills are often underdeveloped in recent college graduates (Hart Research Associates, 2015, 2018; Drummond & Selvaratnam, 2009; Coll & Zegwaard, 2006; Hanson & Overton, 2010; Hill, Overton, Thompson, Kitson, & Coppo, 2019; Kondo & Fair, 2017; Sarkar, Overton, Thompson, & Rayner, 2016). In particular, these reports noted that critical thinking, problem solving, communication, and teamwork should be developed as part of a solid educational foundation across science, technology, engineering, and mathematics (STEM) fields. Likewise, multiple STEM fields have indicated the crucial need to further develop professional skills outside the standard sphere of technical expertise and disciplinary knowledge (ABET Engineering Accreditation Commission, 2016; American Chemical Society Committee on Professional Training, 2015; The Royal Society, 2014).

In chemistry, the Committee on Professional Training of the American Chemical Society (ACS) mirrors the requests seen in international reports by calling upon institutions that seek a certified ACS chemistry degree to maintain programs that develop transferable student skills beyond traditional content knowledge (American Chemical Society Committee on Professional Training, 2015). The ACS guidelines explicitly state that the Committee on Professional Training "expects that approved chemistry programs will have an established process by which they define, impart, and assess the development of student skills throughout the curriculum" in order to "develop critical skills needed for a professional chemist." Among the skills listed as required are problem solving, critical thinking, oral and written communication, and teamwork.

In this chapter, we describe the use of rubrics as assessment and feedback tools to facilitate student development of professional skills in undergraduate chemistry courses that span the chemistry curriculum. Specifically, we report the impact of rubric usage on instructor practices, teaching assistant professional development, and student understanding of professional skills. A broad overview of how the rubrics have been implemented in multiple learning environments will be followed by detailed descriptions of two case studies.

Supporting Success in Active Learning Environments

While preparing STEM students for successful and fulfilling careers after college is an important goal for many institutions of higher education, of equal importance is ensuring that students develop the skills they need to succeed as students in their undergraduate STEM courses. Recent systemic shifts in science education toward active learning and collaborative pedagogies call upon students to develop more than a mastery of content knowledge to be successful (National Research Council, 2012; Kober, 2015). Further, students are being challenged to apply knowledge in new ways, identify innovative strategies for addressing real-world problems, and perform collaborative work that includes internships and course-based undergraduate research experiences. Several active learning strategies, including open-ended laboratory experiments and research projects, are ideal for providing students with opportunities to synthesize and integrate disciplinary knowledge and apply it to real-world settings. Starting in their foundational chemistry courses, students need to develop the professional skills that will help scaffold their learning for ultimate success in advanced chemistry courses, research opportunities, and other high-impact practices.

Explicit development of skills such as communication and teamwork throughout STEM curricula can provide students with tools to be successful in these environments and the confidence in their abilities to further their growth and skill development. Although it has been demonstrated that employers and instructors value professional skills, it is less common for instructors to explicitly assess these skills in their courses (National Research Council, 2012). This lack of constructive alignment between course design and intended outcomes decreases the likelihood that students will make significant gains in these areas.

Constructive Alignment

The theory of constructive alignment is rooted in the constructivist view of learning in which individuals construct their own knowledge rather than receive knowledge from another person (Bodner, 1986). Constructivism posits that multiple factors impact the knowledge that students construct such as their prior knowledge, their social learning environment, and the actions of their instructor and peers (Ausubel, Novak, & Hanesian, 1978; Bodner, 1986). In the constructive alignment model, students will construct the knowledge that an instructor wants them to learn when there is alignment between a course's intended learning outcomes, the tasks that instructors and students perform, and the ways in which students are assessed (Biggs, 1996, 2003, 2014). For example, if an instructor states that "developing critical thinking skills" is a desired learning outcome for the course, then the instructor should be explicit about that desired outcome at the outset of the course, students should engage in tasks that call upon them to use critical thinking, and students should be assessed on their critical thinking and given feedback. All three aspects (intended outcomes, tasks, and assessments) indicate to students what the instructor values and where the students should focus their attention (Biggs, 2003). We have found that instructors who want to be more explicit about assessing their students' professional skills and providing them feedback on their development often lack the tools to do so.

Assessment of Professional Skills

Most resources for assessing professional skills in the classroom are limited to very specific assignments, and typically focus on oral or written presentation skills. Teamwork is also frequently assessed, but often the assessment focus is centered on the relative contributions of each team member to the final product. While the American Association of Colleges and Universities has designed a set of VALUE (Valid Assessment of Learning in Undergraduate Education) rubrics to assess a range of student skills (Association of American Colleges & Universities, 2010), the focus was on programmatic assessment of graduate outcomes rather than to provide feedback to students at the course level. The Enhancing Learning by Improving Process Skills in STEM (ELIPSS) Project (Cole, Lantz, & Ruder, 2016) was established to address the lack of constructive alignment by developing resources to support instructors in assessing development of students' professional skills and providing feedback directly to students about their progress in the development in these skills (Cole, Lantz, Ruder, Reynders, & Stanford, 2018). The skills focused on by the ELIPSS project team emerged from the POGIL (Process Oriented Guided Inquiry Learning) Project: problem solving, critical thinking, information processing, oral/interpersonal and written communication, teamwork, management, and metacognition (Moog, Creegan, Hanson, Spencer, & Straumanis, 2006; Moog & Spencer, 2008). Rubrics to assess both student artifacts and student interactions in classroom or laboratory settings were developed using an iterative approach informed by the literature. Numerous experts and students across the range of STEM disciplines also provided feedback as part of this iterative process (Cole et al., 2018; Ruder, Stanford, & Gandhi, 2018). The rubrics have been implemented over three years at fourteen institutions, and it was determined that the rubrics generated valid and useful assessment data in multiple STEM disciplines (Cole et al., 2018).

An operationalized definition of each skill was drawn from how these skills have been defined in the literature as well as through the consensus of faculty members from multiple STEM fields; these definitions appear at the top of each rubric. An example of an ELIPSS rubric used to assess information processing occurring in student interactions is shown in Fig. 13.1. A score of zero indicates no evidence was observed for a particular category. No descriptors are provided for scores of two and four, which serve to allow assessors more flexibility when student work or interactions have aspects of two performance levels.

Information Processing	Evaluating, interpreting, and manipulating or transforming information.				
Category	**0**	**1**	**2**	**3**	**4**	**5**
(A) Evaluating		A. **Minimally** determined the significance or relevance of information/data needed for the task.		A. **Partially** determined the significance or relevance of information/data needed for the task		A. **Completely** determined the significance or relevance of information/data needed for the task
(B) Interpreting		B. **Inaccurately** provided meaning to data, made inferences and predictions from data, or extracted patterns from data		B. Provided meaning to data, made inferences and predictions from data, or extracted patterns from data **with some errors**		B. **Accurately** provided meaning to data, made inferences and predictions from data, or extracted patterns from data
(C) Manipulating or Transforming		C. **Ineffectively** converted information/data from one form to another		C. **Partially** converted information/data from one form to another		C. **Effectively** converted information/data from one form to another

Fig. 13.1 ELIPSS rubric to assess information processing in student group interactions

Faculty Implementations

One focus of the ELIPSS project has been to study the implementation of the ELIPSS rubrics at multiple institutions and across all undergraduate STEM course levels. The rubrics were designed to be flexible in their use and allow instructors to assess students' professional skills in a variety of learning environments. In addition to providing students with immediate feedback on professional skill development, the ELIPSS rubrics were also intended to provide instructors with feedback on the efficacy of the classroom strategies they employ to develop these skills. To explore the ways in which chemistry faculty implemented the rubrics, chemistry instructors who had indicated they were using the rubrics in the classroom were invited to complete a survey describing their implementation. More detailed case studies were conducted for two of these instructors that included feedback from students and teaching assistants in those courses. Institutional Review Board approval was obtained before commencement of the study (IRB ID#: 201506758).

Participants and Setting

While the rubrics have been implemented by instructors from a variety of STEM disciplines, herein we focus specifically on implementation in chemistry. Seven chemistry professors from seven institutions in the USA incorporated the rubrics into their courses. As shown in Table 13.1, the courses spanned the full range of undergraduate chemistry courses (introductory to advanced) and included both large enrollment courses at research-intensive universities and smaller classes at primarily undergraduate institutions.

Table 13.1 Chemistry courses in which the rubrics were implemented from spring 2016 to spring 2018

Course	Semesters	Institution type	Class size	Professional skills assessed
General Chemistry I—Lecture	Fall 2017	R	50	IP
	Fall 2016, Fall 2017	PUI	23	IC, PS, MC, TW
General Chemistry I—Lab	Fall 2017, Spring 2018	PUI	15	MG, WC
General Chemistry II—Lecture	Spring 2016, Spring 2017	R	150	CT, PS, TW
	Spring 2017	PUI	18	IC, IP
Human Chemistry (non-majors)	Spring 2018	PUI	60	IP
*Organic Chemistry I	Fall 2016, Fall 2017	R	225	CT, IP, PS, TW
*Organic Chemistry II	Spring 2017, Spring 2018	R	150	CT, IP, PS, TW
*Physical Chemistry I—Thermodynamics	Spring 2017, Fall 2017	R	30	CT, IC, IP, PS, TW
Integrated Science Seminar (upper-level)	Fall 2016, Fall 2017	PUI	20	CT, IC, MG, TW
Instrumental Analysis	Spring 2016, Spring 2017, Spring 2018	PUI	22	CT, IC, PS, TW

*Indicates courses for which case studies are described in more detail below. Institution type: R = research-intensive university, PUI = primary undergraduate institution. Professional skills assessed: CT = critical thinking, IC = interpersonal (oral) communication, IP = information processing, MC = metacognition, MG = management, PS = problem solving, TW = teamwork, WC = written communication

Data Collection

Participating chemistry instructors used the ELIPSS rubrics to assess their students' written work and group interactions over a three-year period. These instructors provided feedback about their experience using the rubrics to the research team via surveys at the end of each semester. The surveys requested demographic information such as course name, class size, and which rubrics were used (Table 13.1). The surveys also included open-response questions that prompted the instructors to describe the student assignments that they scored with the rubrics, why the rubrics were appropriate to assess those assignments, how the instructors gave feedback to students, and how students responded to the feedback.

For the two case studies, additional data were collected in the form of interviews and undergraduate teaching assistants' (UTAs) written reflections. Twenty UTAs from the organic chemistry course (Case Study 1) wrote two reflections each semester about facilitation, their understanding of professional skills, their ability to assess professional skills, and the effectiveness of different assessment strategies. Six of these UTAs were interviewed over the two years of the study. In these semi-structured interviews, UTAs discussed what behaviors they looked for as evidence of different skills when assessing students in the classroom, how the training they received made them more aware of professional skills, and how they applied this training to other situations. Undergraduate students from the physical chemistry course (Case Study 2) were interviewed at the end of the semester to determine their perceptions of how the rubrics were implemented in the course, how well they thought that the rubric scores reflected their work, how they used the feedback to improve, and how the actions of the teaching assistants and instructor affected the students' understanding of the rubric feedback. Three students volunteered to participate in semi-structured interviews. Additionally, the two graduate teaching assistants (GTAs) for the course were interviewed at the end of the semester to determine how they used the rubrics to assess the students' written work and group interactions, how they provided feedback to the students, and how using the rubrics affected their perceptions of the course. These interviews were also semi-structured.

Data Analysis

The faculty responses to the survey questions were analyzed qualitatively. Open-coding was used to develop an initial coding scheme; then the codes were refined over multiple iterations and grouped to form themes related to rubric implementation and methods of providing feedback. In addition to the faculty data, the student and teaching assistant interviews from Case 1 and Case 2 were analyzed to develop a separate set of themes. The interviews were transcribed verbatim, and the interview transcripts and TA reflections were qualitatively analyzed using a grounded theory approach and constant comparative method (Strauss & Corbin, 1990). Open-coding was again used to develop an initial set of codes, and the codes were grouped into themes that represented various aspects of the participants' views of the professional skill assessment in their courses.

Results and Discussion

Overall, the survey data from faculty participants indicated that rubrics were implemented in a range of chemistry courses (Table 13.1) and rubric use was tailored to meet the needs of individual courses and learning environments. A summary of the themes identified from the analysis of the collected implementation data can be found in Table 13.2.

Rubric Format

The way in which instructors used the rubrics varied across implementations. Most instructors used the entire rubric to assess their students and provide feedback for at least one professional skill. A few instructors also used one or two categories from a rubric (i.e., a partial rubric) at a time to have students focus on a particular aspect of a skill for some assessment tasks. In some cases, instructors combined categories from different ELIPSS rubrics to create their own composite rubrics. Instructors often combined categories from the problem solving, critical thinking, and information processing rubrics to align with particular student tasks. We argue that students cannot engage in problem solving or critical thinking without first using their information processing skills, so it makes sense that instructors would assess these skills in tandem. Additionally, we found that instructors incorporated ELIPSS rubric categories into their existing grading rubrics when creating these composite rubrics. One instructor reported that "the critical thinking rubric was incorporated into the rubric I used for grading on… our course management system." This was common when instructors were assessing laboratory reports, allowing the instructors to more closely align their assessments with their intended learning outcomes (both in terms of content and professional skill development) for a particular task. Although the creation of composite rubrics was not an original intention of the ELIPSS project, this is seen as a positive sign that instructors are incorporating professional skill assessment and feedback into their curricula. The fact that the rubric implementation was amenable to variations other than using the rubrics "as written" indicates that the rubrics are consistent with the change strategy of developing reflective teachers (Borrego & Henderson, 2014) in which innovations (i.e., the ELIPSS rubrics) can be modified by instructors to fit their individual environments and goals.

Provider of Feedback and Audience for Feedback

In all cases, the instructor for the course provided feedback to students on their professional skill development in some form. The most common method of delivering feedback to students was to summarize observed performance, along with desired outcomes, aloud to the entire class. Methods of providing specific feedback to learning teams and individual students were more varied. Instructors of large-lecture courses noted that it was difficult to frequently give feedback to groups and individuals. They often addressed this difficulty by having teaching assistants provide

Table 13.2 Themes identified from the analysis of rubric implementation data

Themes	Codes	Code definitions
Rubric format	Entire	All categories (rows) from a single ELIPSS rubric were used
	Partial	One or two categories (rows) from a single ELIPSS rubric were used
	Composite	Categories (rows) from multiple ELIPSS rubrics were combined to make a customized rubric
Provider of feedback	Instructor	The course instructor analyzed student work or group interactions
	Teaching assistants	A graduate or undergraduate teaching assistant analyzed student work or group interactions
	Peers	Students analyzed the work of other students in the class or the interactions of members of their group
Audience for feedback	Whole class	The instructor discussed the rubrics and/or professional skills with the entire class
	Learning team/group	Scores and written comments on the rubrics were provided based on group interactions or written work
	Individual students	Scores and written comments on the rubrics were provided based on individual written work
Nature of feedback	Completed rubric	A written assignment was scored with a rubric; then the rubric was returned to the student along with the assignment. Group interactions were scored using a rubric and the completed rubric was given to the group
	Completed rubric with explanatory comments	A written assignment was scored with a rubric along with added comments to justify scores; then the rubric was returned to the student along with the assignment. Group interactions were scored using a rubric with added comments to justify scores and the completed rubric with comments was given to the group
	Description of specific behaviors/evidence	Students were provided a list of behaviors and/or evidence for professional skill(s) observed on an assignment or during group work
Strategies for encouraging student reflection	Use of the same rubric(s) on multiple assignments	The same rubric was used to assess students multiple times throughout a semester
	Feedback followed by revision	Students were given feedback on an assignment and then allowed to make edits before resubmitting the same assignment
Reflective practice	Clear assessment criteria	Instructors and teaching assistants were able to clearly articulate their assessment criteria for themselves and the students
	Greater awareness of professional skills in group interactions	Instructors were more aware of how groups interacted and aspects of professional skills that needed to be addressed

feedback to students. In a few cases, instructors had students provide feedback to one another (i.e., peer review).

The audience for feedback varied by the number of students who received feedback simultaneously. As mentioned above, delivering general comments to the entire class was an easy and efficient way to provide feedback about observed skills to all students. Additionally, instructors or their teaching assistants often provided more specific feedback to learning teams or individual students. For example, one instructor noted that she gave feedback "in the form of comments to the class and personal comments on specific journal entries" after the students had filled out journal entries in which they reflected on their skills. It was most common to provide feedback to individual students on written work and to learning teams on their interactions.

Nature of Feedback

After assessing students with the ELIPSS professional skill rubrics, most instructors simply gave the scored rubric back to the students. When assessing written work, the rubric was returned with the individual student's assignment. When assessing interactions, however, only the scored rubric was provided to the students because there was no artifact to accompany it. Some instructors and teaching assistants provided written comments on the rubrics that either explained the students' scores or provided suggestions for improvement. Additionally, instructors often provided students with a list of expected behaviors that would act as evidence that they engaged in the intended professional skills or observations of different behaviors observed for a particular skill. For example, students were instructed to "direct each other to tasks/information" and "acknowledge statements of other team members" when engaging in teamwork. In a large classroom, the list of desired behaviors was delivered as a "clicker" question and students were asked to report which behavior(s) they had engaged in. The student responses served as the foundation for further discussion of what was observed by the instructor.

Strategies for Encouraging Student Reflection

The most common approach used by instructors to encourage students to consider how they could improve their skills was to use the same rubric or rubrics to assess students at multiple times throughout the semester. Students were encouraged to reflect on the feedback provided and plan for how they could improve their performance on subsequent assignments. This process occurred with the entire, partial, and composite rubrics. An alternate strategy was employed by one instructor for a general chemistry laboratory course who had students assess each other on an initial draft of their laboratory report and then refine the report before resubmitting it for their final grades. According to the instructor, "Students were given feedback via peer review and from the TA (graduate assistant). Following the peer review process, students were allowed to rewrite their reports. We saw significant improvement from the first draft to the second. Then, [the students] were assessed with the same composite rubric on another lab report nearer the end of the semester. We saw significant improvement from mid- to end-of-semester as well." These facilitation techniques are in line with the theory of constructive alignment by having students engage in tasks (i.e., reflecting on professional skills) that may help students achieve the intended learning outcome (i.e., developing professional skills).

Reflective Practice

In regard to professional development, instructors reported that after implementing rubrics, they were generally more cognizant of their students' professional skills, particularly when using interaction rubrics during in-class group work. Reflecting on student performance helped instructors clarify and articulate their expectations for themselves and for students. Teaching assistants from the physical chemistry course mirrored the observations of the faculty by stating that the rubrics clarified which aspects of student skills they should assess. The instructors found that spending some class time going over the rubrics helped clarify the expectations of skills needed to succeed in an active learning environment. For example, when observing student groups, instructors were more likely to notice which groups had one student dominate conversations and could use the rubrics to give feedback to groups to encourage participation among all group members.

Case studies 1 and 2 below are illustrative examples of these implementation themes by demonstrating how faculty members used the rubrics to assess and provide feedback to students on the development of their professional skills. Moreover, these cases provide insight into how students responded to the feedback.

Case 1: Organic Chemistry

The ELIPSS rubrics were implemented in a large enrollment organic chemistry course at a large urban, research-intensive university from fall 2016 through spring 2018. There were approximately 150–225 students enrolled in the course each of the four semesters, with predominately second- and third-year undergraduate students majoring in chemistry, biology, and pre-health-related areas. The course was facilitated by the same experienced instructor and 8–10 undergraduate teaching assistants (UTAs). The course was taught using POGIL materials (Ruder, 2015) in a fixed-seat auditorium where students spent a majority of class time working through the POGIL activities in structured learning teams that had assigned roles. Students regularly completed group quizzes that were graded and assessed for evidence of professional skills. The professional skills assessed during the semester were teamwork, information processing, critical thinking, and problem solving.

Rubric Implementation

As part of regular classroom practice, a combination of partial, entire, and composite rubrics was used. When introducing a new skill such as information processing, a single category of the rubric (i.e., partial rubric) was used. After the UTAs and students gained experience using the individual categories, the entire rubric was used to assess group interactions. Terms from the relevant ELIPSS rubrics were incorporated into the grading rubrics used to assess group quizzes. Feedback came from multiple sources: the instructor, UTAs, and peers. The instructor graded the group quizzes and used rubric scores from the UTAs and her own observations to provide feedback to students in whole class discussions. The UTAs assessed learning teams with the ELIPSS interaction rubrics while teams were completing group quizzes or working on activities during class. Student learning teams engaged in peer discussion to assess their group performance and assigned themselves a score using the rubrics. In this implementation, feedback was provided to the whole class and to learning teams but was not provided to individual students.

UTAs used rubrics each class period to assess student interactions. Each rubric was used multiple times, with the entire rubric being used at least 2–3 times for each skill. During the in-class group work, UTAs interacted with each group of students, observed how they were engaging in a specific skill, and assessed their interactions using a rubric. When students asked UTAs questions about a concept, instead of answering the question directly, the UTAs posed a series of questions to help scaffold students' learning and engage in the targeted professional skill to arrive at the answer to the question. The UTAs were instructed to use the language from the rubrics to reinforce when students were engaging in a professional skill. They provided informal feedback to groups during class and recorded scores on a sheet that was given to the instructor at the end of class.

Each learning team completed a reflection report at the end of each class period in which they were asked to reflect on the content and skills used that day. At the end of the quiz, student groups were asked to rate their performance on one aspect of the targeted skill and then justify their score. An example of one of these prompts is shown in Fig. 13.2. The group assessments of professional skills typically employed the use of individual categories from the rubrics, but each learning team completed an entire rubric for each skill at some point in the semester.

The feedback from the instructor typically blended the professional skills with their use in applying knowledge to solve problems. She purposefully reiterated language from the rubric to highlight an aspect of a specific professional skill when working through an organic chemistry problem. For instance, when emphasizing information processing while working through an example, the instructor verbalized her actions and wrote the terms evaluating, interpreting, and transforming next to the steps of the example when these skills were utilized. The rubric used for grading

Evaluate your group's ability to form an argument to answer #1b by circling the appropriate number. Justify your choice.

Critical Thinking: Analyzing, evaluating, or synthesizing relevant information to form an argument or reach a conclusion supported with evidence.

Category	0	1	2	3	4	5
Forms and Supports Arguments		Forms **ineffective arguments**: makes claims and provides **little evidence** to support them		Forms **adequate arguments**: makes claims and provides **some evidence** to support them		Forms **effective arguments**: makes claims and provides **complete evidence** to support them

Justification:

Fig. 13.2 Example of a student reflection task following a group quiz

group quizzes also provided specific examples of how the skills should have been applied in the specific context. The repeated use of each rubric and the emphasis of skills by the instructor as part of the grading scheme encouraged students to reflect on the skills and improve performance.

The nature of weekly meetings with the UTAs promoted reflective practice by the instructional team. During these meetings, the instructor asked for comments regarding the successes and challenges the UTAs faced the previous week in using the rubrics, assessing the development of skills, providing feedback, and how students received and responded to the feedback. As part of these discussions, the instructor and UTAs considered changes that could be made to classroom facilitation to help address any challenges they had with assessing and providing feedback. The instructor prompted UTAs to reflect on how the training they received prepared them to identify and assess professional skills and the overall effectiveness of each variation in implementing the rubric and providing feedback.

Insights

During the organic chemistry course, the teaching assistants' process of assessing the students and then reflecting on their assessment practices led to a number of insights. The UTAs found that it was critical for them to work through the definitions of the professional skills and rubric categories to help become familiar with the skills they would be evaluating. During their weekly meetings, the UTAs completed organic chemistry problems and then discussed how each category of the targeted professional skill was used to solve the problem. Finally, the UTAs discussed how evidence of this skill could potentially be seen in student behaviors or student work. This detailed practice began after a UTA commented that, "Information processing was a new skill for me. Critical thinking and problem solving I had heard of before, but they were very vague terms. By going through the rubrics and definitions it helped me understand what people mean when they say those terms and it brought a lot more meaning to those words."

UTAs found that the training focused on discussions of professional skills was helpful in learning how to assist students during group work. One UTA commented that the training was useful because "by knowing that a problem was using [the Analysis category] from [the] critical thinking [rubric], I was able to come up with guiding questions to lead students to a certain point." After observing the groups working during class, UTAs scored a rubric for each group, and this feedback was returned to each group. Initially, students were not graded based on their professional skill performance and were instead given formative and summative feedback solely based on the UTAs observations and interactions. In general, this early attempt at summative feedback of the student group interactions resulted in two reactions from the students. The first reaction was indifference; students would quickly glance at the feedback and put it away. The second was defensiveness; students did not think that they should be formally assessed on anything other than content knowledge, particularly from the UTAs. To help overcome these issues, the development of professional skills became a component of the course grade as part of the students' group quizzes.

In order to receive credit for their rubric ratings, the students were required to provide evidence of the actions they took in using the targeted skill to accomplish their task. The instructor then graded the student justifications to determine how well students could support their rating. It should be noted that even if students rated themselves with the highest score on the rubric, they did not receive full credit if their justification did not support their self-assessment. When the quiz was returned to the students, they received written comments from the instructor on their self-assessment justification as well as rubric scores from UTAs observations. This combined feedback allowed for comparison between what the UTA observed versus the group's self-assessment of how well they engaged in a specific skill. Self-reflection from group members on group quizzes improved as the semester went on and more feedback was provided. For example, students' early reflection to justify how they used evidence to support their claim on the critical thinking rubric used some of the words from the critical thinking definition without describing what they did: "We were able to form a good argument but we sometimes lack the knowledge to come to the correct conclusion." A later reflection was more descriptive about using specific evidence: "We were able to form an argument (for question 1b) by gathering our knowledge on electrophiles and concluding that electrophiles should have a good leaving group while nucleophiles should have a negative charge."

Case 2: Physical Chemistry I—Thermodynamics

In the fall of 2017, the rubrics were used in a physical chemistry thermodynamics course at a large, research-intensive university. This course was required for chemistry majors and was taught in an active learning classroom using POGIL materials (Shepherd, Grushow, Hunnicutt, & Moog, 2017). The course enrollment consisted of 30 upper-level chemistry and biochemistry students and was facilitated by an experienced chemistry instructor and two chemistry graduate teaching assistants (GTAs). Before the course began, the GTAs received training to use the rubrics by participating in an ELIPSS workshop that was run by a member of the research team. Throughout the course, students worked in structured teams with assigned roles to complete POGIL activities during each class, and each team submitted a reflection report at the end of each class.

Rubric Implementation

A combination of partial and entire rubrics was used to assess oral communication, teamwork, information processing, and critical thinking in group interactions. Entire rubrics were used to assess information processing, problem solving, and critical thinking in student written work. The reflection reports called upon students to reflect on the chemistry content they learned that day and assess their group interactions using a single category from one professional skill rubric. Learning teams self-assessed using the entire rubrics (rather than a single category) periodically throughout the semester. Each time the students assessed their teams with an entire rubric, the GTAs also assessed group interactions during class using the same rubric. Students also submitted individual problem sets for grading each week. The GTAs graded the problem sets and periodically assessed each student's written work using an entire rubric in addition to scoring the work for content accuracy.

Feedback was primarily provided to students through completed rubrics. TA scores and comments from the interaction rubrics were added to the group assessment to provide students with a way to calibrate themselves. Completed rubrics and comments were appended to student work when they were applied to problem sets. The instructor provided general feedback about desired behaviors during whole class discussion and indicated areas where learning teams needed to improve their performance.

Insights

Students were interviewed at the end of the course about how they were assessed with the rubrics. During these interviews, students reported that they improved their awareness of the role of developing professional skills in their courses and offered insights that can improve future rubric implementations. As previously mentioned, the students assessed themselves as a group using one rubric category at the end of each class as part of their daily reflection reports. However, students reported that they did not have adequate time for self-reflection at the end of each class period. One student noted that "Everybody rushes through [the reflection reports] cause we're always short on time" and indicated that this was a widespread issue and not limited to one group. The interviewed students were each in different groups and reported the same issue. In addition to a lack of time for self-assessment, students also reported a lack of incentive to carefully assess themselves. The instructor for the course did write comments on the reflection reports when students wrote inadequate or inaccurate self-assessments, but the students noted that making the reports worth points toward their final grade would encourage them to spend more time filling out the report. As one student said, "if you put points on anything, people will definitely put more attention to it."

When the students were asked about the feedback they received on their written work, they noted that they did not see the rubrics for written work before they submitted their problem sets, so they did not know how they would be assessed. They only saw the scored rubrics that they were given when the problem set was returned to them. One of the main tenets of constructive alignment is that students should have a clear idea of the desired learning outcome for a task. Since the students did not see the rubrics before they completed the problem set, they may not have had a clear idea of the skills that they were meant to practice and develop when they completed the task, and this may have led to a misalignment between the students' perception of the task and the instructor's intended learning outcomes. In regard to using the feedback they received, all three interviewees stated that they needed clearer connections between the course content and the professional skills if they were going to be able to use the rubric feedback in a productive manner. Specifically, they said they needed clearly articulated ways to improve, with one student suggesting that the instructor or TA should "Try to point out what I should do to… get a five [the highest rubric score]."

From the interviews of the graduate teaching assistants in the thermodynamics course, it was found that the rubrics helped them to define assessment criteria and to determine the degree to which students were achieving the intended learning outcomes. When referring to scoring student written work, one GTA said, "I found the rubrics helpful for me in order to [better understand] how I should be analyzing the students, how I can understand if they're getting what they should out of it." Later, when referring to assessing groups of students with the interaction rubrics, the same GTA said, "I thought that it was helpful for me to understand how my students should be interacting, what it is that I should be looking for." This GTA's reflections demonstrate the opportunity to use the ELIPSS rubrics not only to help faculty members become more reflective practitioners, but also to do the same for graduate teaching assistants.

Case Study Analysis

The two case studies showed similar results regarding student perceptions of the professional skill feedback. In both cases, the instructor discussed the importance of professional skills at the beginning of the course to frame their students' perceptions, yet students in both cases initially reported general indifference toward using the feedback they received because they felt inadequately incentivized to meaningfully engage with it. However, when the students from Case 1 were given grade incentives in the latter part of their course, they took it more seriously, as evidenced in their responses to self-assessment. The students from Case 2 did not receive this same incentive, but they did report in their interviews that they would be more likely to act on the feedback if they did receive such an incentive. Both cases also demonstrated that students need concrete examples of how to connect the professional skills and their course content.

The major difference between the two cases was the pushback against professional skill assessment that was seen in Case 1. Although we cannot definitely state why this difference occurred, a number of factors could contribute to this disparity. The courses had different class sizes and course levels and were taught at two different institutions. While the organic chemistry courses involved large class sizes of 150–225 and acted as a service course that included second- and third-year pre-health students, the thermodynamics course was much smaller with 30 students that were predominately third- and fourth-year chemistry and engineering majors. These different demographics could have had an effect on the student perceptions of the professional skill feedback they received.

For the undergraduate and graduate teaching assistants from the two cases, the training they received aided their professional development by making them better able to assess student work and facilitate active learning group work. The organic chemistry UTAs were trained to use the professional skill rubrics during weekly meetings by discussing professional skills and behaviors expected, while the thermodynamics GTAs were trained using ELIPSS workshops that were run by a member of the research team. Both sets of TAs reported that the specific scoring criteria on the rubrics helped clarify the course instructors' expectations for what the students should be achieving and helped them give feedback to students. In order to achieve constructive alignment, it is important that an instructor's actions and facilitation choices reflect the outcomes they want students to achieve (i.e., developing professional skills). Since the TAs often had more face time with the students than the course instructor, making the instructor's expectations clear to the TAs may help the TAs make those expectations clearer for the students.

Conclusions

A number of general themes emerged from the analysis of the faculty implementations in all chemistry courses and feedback from the students and TAs from the two case studies. Rubric implementation surveys completed by participating instructors indicated that they had developed implementation strategies that suited their intended learning outcomes by assessing and providing feedback to students in a variety of ways. Instructors with different class sizes, course levels, online course management systems, and access to teaching assistants all adapted the rubric implementation strategy to fit their unique classroom environments. The instructors used entire, partial, and composite rubrics to tailor their assessments to student tasks. Feedback was provided by instructors, teaching assistants, and peers. Feedback was provided to whole classes, learning teams/groups, and individual students. The feedback was provided in the form of completed/scored ELIPSS rubrics that were sometimes accompanied by explanatory comments or descriptions of specific behaviors that the instructors wanted to see in the student groups. While the instructors encouraged the students to reflect on their professional skills, the instructors and teaching assistants also engaged in reflective practice by reflecting on their assessment strategies. Multiple instructors reported that they were better able to articulate professional skill expectations to their students through the use of the rubrics. Additionally, they were more aware of how their students interacted with one another in groups after using the interaction rubrics. We see this as evidence that the professional skill rubrics can encourage more reflective practice in undergraduate instructors by providing them with more information about their students that can be used to modify their teaching methods.

Implications for Practice

Through the analysis of the surveys that indicated how rubrics were used as part of a strategy to develop and assess professional skills, we developed the scheme shown in Fig. 13.3 to illustrate how the use of ELIPSS rubrics provided important feedback mechanisms to the instructors and the students.

The first step is choosing which professional skills will be assessed during the course and in which order they are introduced. The instructor must then choose course materials and classroom facilitation strategies that provide students opportunities to develop and display evidence of their skills while completing tasks or engaging in

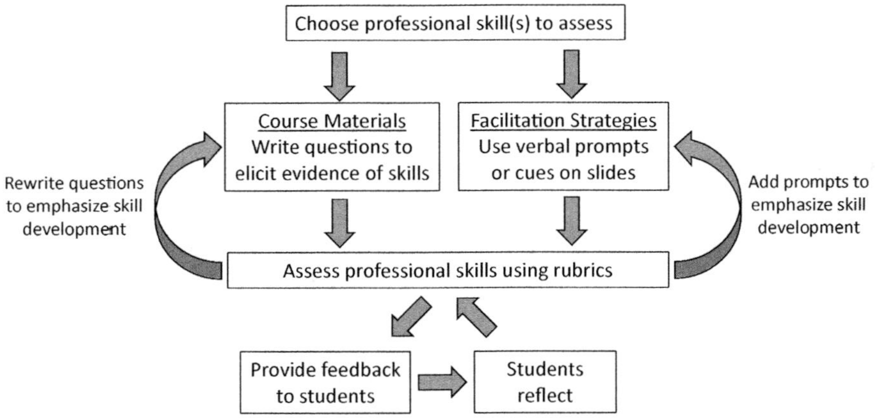

Fig. 13.3 Flowchart of how ELIPSS rubrics can be used to inform reflective practice for both instructors and students

group discussions. This may require rewriting questions to direct students to show evidence of their professional skills in their written responses or adding explicit prompts to slides or use other facilitation strategies to encourage group interactions. Next, instructors should use the rubrics to assess student written work and group interactions to provide themselves with insights as to the degree to which students have successfully displayed the targeted skill. Analysis of student performance provides data to indicate areas for further development (which benefits both students and instructors) and may indicate areas where materials and/or classroom facilitation can be improved. Providing feedback to students and having them reflect on the feedback from the instructional team also allow students to identify areas for improvement. Intentional use of strategies to develop and assess professional skills in student interactions and written work results in improved instructional choices on the part of instructors and TAs and improved performance from students.

Acknowledgements This work was supported in part by the National Science Foundation under collaborative grants #1524399, #1524936, and #1524965. Any opinions, findings, and conclusions or recommendations expressed in this material are those of the authors and do not necessarily reflect the views of the National Science Foundation. We thank members of our Primary Collaboration Team and Implementation Cohorts for collecting and sharing data. We also thank all the students who have allowed us to examine their work and provided feedback.

References

ABET Engineering Accreditation Commission. (2016). Criteria for accrediting engineering programs. http://www.abet.org/accreditation/accreditation-criteria/criteria-for-accrediting-engineering-programs-2016-2017/. Accessed February 2017.

American Chemical Society Committee on Professional Training. (2015). *Undergraduate professional education in chemistry: ACS guidelines and evaluation procedures for bachelor's degree programs*. Washington, DC: American Chemical Society. https://www.acs.org/content/dam/acsorg/about/governance/committees/training/2015-acs-guidelines-for-bachelors-degree-programs.pdf. Accessed December 2018.

Association of American Colleges & Universities. (2010). VALUE rubric development project. https://www.aacu.org/value/rubrics. Accessed February 2017.

Ausubel, D. P., Novak, J. D., & Hanesian, H. (1978). *Educational psychology: A cognitive view*. New York: Holt, Rinehart, Winston.

Biggs, J. (1996). Enhancing teaching through constructive alignment. *Higher Education, 32*(3), 347–364.

Biggs, J. (2003, April). Aligning teaching and assessing to course objectives. *Teaching and learning in higher education: New trends and innovations, 2,* 13–17.

Biggs, J. (2014). Constructive alignment in university teaching. *HERDSA Review of Higher Education, 1*(1), 5–22.

Bodner, G. M. (1986). Constructivism: A theory of knowledge. *Journal of Chemical Education, 63*(10), 873–878.

Borrego, M., & Henderson, C. (2014). Increasing the use of evidence-based teaching in STEM higher education: A comparison of eight change strategies. *Journal of Engineering Education, 103*(2), 220–252.

Cole, R., Lantz, J., & Ruder, S. (2016). ELIPSS: Enhancing Learning by Improving Process Skills in STEM. http://elipss.com/. Accessed February 2017.

Cole, R., Lantz, J., Ruder, S., Reynders, G. J., & Stanford, C. (2018, June). *Enhancing learning by assessing more than content knowledge*. Paper presented at the 2018 ASEE Annual Conference & Exposition, Salt Lake City, Utah.

Coll, R. K., & Zegwaard, K. E. (2006). Perceptions of desirable graduate competencies for science and technology new graduates. *Research in Science & Technological Education, 24*(1), 29–58.

Drummond, H. P., & Selvaratnam, M. (2009). Intellectual skills needed for the effective learning and application of chemical knowledge. *South African Journal of Chemistry, 62*(1), 179–184.

Hanson, S., & Overton, T. (2010). Skills required by chemistry graduates and their development in degree programmes. Higher Education Academy, UK Physical Sciences Centre. http://www.rsc.org/learn-chemistry/resources/business-skills-and-commercial-awareness-for-chemists/docs/skillsdoc1.pdf. Accessed December 2018.

Hart Research Associates. (2015). Falling short? College learning and career success. https://www.aacu.org/sites/default/files/files/LEAP/2015employerstudentsurvey.pdf. Accessed December 2018.

Hart Research Associates. (2018). *Fulfilling the American dream: Liberal education and the future of work* (p. 21). https://www.aacu.org/sites/default/files/files/LEAP/2018EmployerResearchReport.pdf. Accessed December 2018.

Hill, M. A., Overton, T. L., Thompson, C. D., Kitson, R. R. A., & Coppo, P. (2019). Undergraduate recognition of curriculum-related skill development and the skills employers are seeking. *Chemistry Education Research and Practice*. https://doi.org/10.1039/c8rp00105g.

Kober, N. (2015). *Reaching students: What research says about effective instruction in undergraduate science and engineering*. Board on Science Education, Division of Behavioral and Social Sciences and Education. Washington, DC: The National Academies Press.

Kondo, A. E., & Fair, J. D. (2017). Insight into the chemistry skills gap: The duality between expected and desired skills. *Journal of Chemical Education, 94*(3), 304–310.

Moog, R. S., Creegan, F. J., Hanson, D. M., Spencer, J. N., & Straumanis, A. R. (2006). Process-oriented guided inquiry learning: POGIL and the POGIL project. *Metropolitan Universities, 17*(4), 41–52.

Moog, R. S., & Spencer, J. N. (2008). *Process oriented guided inquiry learning (POGIL)*. Washington, DC: ACS Publications.

National Research Council. (2012). *Discipline-based education research: Understanding and improving learning in undergraduate science and engineering*. Washington, DC: The National Academies Press.

Ruder, S. (2015). *Organic chemistry: A guided inquiry*. New York: Wiley.

Ruder, S., Stanford, C., & Gandhi, A. (2018). Scaffolding STEM classrooms to integrate key workplace skills: Development of resources for active learning environments. *Journal of College Science Teaching, 47*(5), 29–35.

Sarkar, M., Overton, T., Thompson, C., & Rayner, G. (2016). Graduate employability: Views of recent science graduates and employers. *International Journal of Innovation in Science and Mathematics Education, 24*(3), 31–48.

Shepherd, T., Grushow, A., Hunnicutt, S. S., & Moog, R. S. (2017) Thermodynamics & statistical mechanics: A guided inquiry. Unpublished educational materials.

Strauss, A., & Corbin, J. M. (1990). *Basics of qualitative research: Grounded theory procedures and techniques*. Thousand Oaks, CA: Sage Publications Inc.

The Royal Society. (2014). *Vision for science and mathematics education*. London: The Royal Society Science Policy Centre. https://royalsociety.org/~/media/education/policy/vision/reports/vision-full-report-20140625.pdf. Accessed December 2018.

Chapter 14
Students' Perceptions of Group Communication Skills in an Active Learning Environment

Alexandra Yeung and Suzanne Ahern

Abstract Curtin University uses an active learning framework in its first-year chemistry units to help students engage better with content, compared with traditional didactic lectures. It is intended that students develop other skills such as communication. However, these skills are not always explicitly taught and we assume students are proficient communicators when they enter the classroom. This study aimed to create activities targeting development of group communication. An examination of how students communicate with each other within their group as they completed their learning activities was conducted. The Talk Science Primer was used as a framework for the learning design of the activities where students work through four goals to develop communication skills (Michaels and O'Connor in Talk science primer. TERC, Cambridge, 2012). Students ($n = 452$) enrolled in an introductory-level chemistry unit who had very little chemistry background were divided into two groups—intervention and non-intervention. The interventions were activities conducted in class throughout the semester. Students' perceptions of the interventions were determined, particularly the usefulness of the interventions in developing their communication skills.

Introduction

During their education, students learn essential skills such as time management, teamwork and communication. Students learn these skills through both direct teaching and indirect teaching. Teamwork is indirectly developed through experiences such as working in groups and developed by direct instruction that can be applied to other situations (Prince, 2004). Similarly, communication is developed through conversing in the classroom and socialising outside of class. There are many ways to learn how to communicate although most are learnt indirectly through conversations

A. Yeung (✉) · S. Ahern
Curtin University, Perth, Australia
e-mail: alexandra.yeung@curtin.edu.au

S. Ahern
e-mail: suzannebahern@gmail.com

M. Schultz et al. (eds.), *Research and Practice in Chemistry Education*,
https://doi.org/10.1007/978-981-13-6998-8_14

and are not structured. Some structured instruction on communication does occur when learning a new topic to master the discourse that surrounds the new topic. However, direct instruction of communication does not occur past secondary-level education once mandatory English classes cease.

A large portion of classroom culture is related to communication. The way students communicate in class and outside of class affects the way students learn. Some of this communication can enhance learning, while others can be counter-productive. The way teachers communicate to students impacts classroom dynamics, which is a factor in student learning. Teacher communication to students relates to many areas of current research such as teaching pedagogies, teacher–student interaction, student–student interaction and additional resource materials for students to use in their study.

Background

Theoretical Framework: Constructivism

Constructivism is currently regarded as the dominant paradigm for teaching and learning in education. The theory emerged from the work of Kant (Phillips, 1995) and is derived from the work of Dewey (1916), Piaget (1952) and Vygotsky (1978). It is based on the notion that learning is an active process of 'sense-making' (Brown, 1994; Driver, Asoko, Leach, Scott, & Mortimer, 1994; Phillips, 1995). Teachers must apply it effectively for students to create new meanings when information is presented to them (Powell & Kalina, 2009). Constructivism comes in the forms of cognitive constructivism, where ideas are constructed in a personal process, and social constructivism, where ideas are constructed through interaction. Both forms of constructivism rely on clear communication in order to be effective (Phillips, 1995; Powell & Kalina, 2009).

Cognitive constructivists such as Piaget (1952) and von Glasersfeld (1993) view learning as a predominantly individual process where an individual develops his or her own theories. There are also distinct stages of learning, and the last stage of learning is assimilation and accommodation that requires students to construct their new knowledge and resolve conflicts with previously held knowledge. The stages of learning occur at different paces for different students, which impacts on lesson progression as students are most commonly at different places (Powell & Kalina, 2009).

Piaget believed that when the individual constructs knowledge it is based on his or hers personal experience. Learning is influenced by prior knowledge, beliefs, expectations, motivation, perceptions and social and cultural contexts (Alexander, 2007; Bodner, 1986; von Glasersfeld, 1993). As students learn, they learn based upon what they already know. This process of constructing knowledge based on what the learner previously knew is very different to other theories and conflicting paradigms (Powell & Kalina, 2009).

The learning cycle is based on constructivist principles that constitute the closest way students learn (Spencer, 1999). The learning cycle comes in three phases: (1) exploration phase where patterns and hypothesis are sought and generated from data, (2) concept invention phase where concepts are developed and terms are applied and (3) application phase where the newly developed concept is applied to a new situation (Moog & Spencer, 2008; Spencer, 1999). The second phase has been shown to be the most crucial phase when learning new content, and this is aided by social constructivism (Spencer, 1999).

Social Constructivism

More recently, social constructivism and sociocultural theory have become the more widely adopted constructivist positions in educational psychology. These approaches are derived from the work of Vygotsky (1978) and view learning as fundamentally social. Social constructivism is based on the social interactions a student has combined with their personal critical thinking process (Powell & Kalina, 2009). The social interaction leads to a more effective internalisation of knowledge despite the different rates that this occurs in students. Learning is mediated by social factors such as interactions with peers and teachers (Brown, 1994), and a 'scaffold' can be constructed when learners of similar cognitive abilities co-construct understanding of new material (Palincsar, 1998). Discussion between students promotes critical thinking, and students will construct personal meaning that was developed on their own. The language students' use is the most important factor in developing effective communication skills, and social interaction is the best solution for this process (Powell & Kalina, 2009).

The zone of proximal development, or ZPD, is one of Vygotsky's main theories that controls how a student learns. When a student is learning, the learning only occurs within a boundary or zone and without the help of others. Their boundary will not expand to accommodate more learning. The teacher and interaction with other students will allow the ZPD to expand for continued learning (Vygotsky, 1978; Powell & Kalina, 2009). Furthermore, social constructivist perspectives consider learning as a two-step process, firstly on a social level through a process of enculturation into a community by participating in activities (Mason, 2007) and, secondly, within the learner through a process of internalising social and cultural activities (John-Steiner & Mahn, 1996). Social constructivist theories of learning emphasise the role of language, cultural artefacts and the enculturation process already mentioned (Palincsar, 1998). Since learners are enculturated into the social practices of a community, they can be considered apprentices to a community of practice.

One overarching theme of the emerging pedagogies is the utilisation of social constructivism to create knowledge in groups. The sharing and creation of knowledge through social means such as working in groups lead to more student retention and lower fail rates in class (Prince, 2004). These students can achieve a deeper understanding of the content information rather than surface learning. The constructivist nature means that collaboration in groups engages more students and uses their social nature to learn content. The collaborative environment also provides an opportunity for students to engage in other scientific practices such as argumentation (Cole, Becker, & Stanford, 2014).

Active Learning

The dominant teaching method at tertiary institutions has been through lectures dictated by academic staff to students (Spencer, 1999). In this didactic teaching method, students are generally not engaged or motivated and have a tendency to passively learn, and student engagement tends to drop drastically after 15 min of a lecture (Prince, 2004). Furthermore, students' performance varies greatly using this instruction type and can lead to only superficial engagement with the content. In recent years, other styles of teaching have been emerging that appeals to more students and promotes deeper engagement with the content using social interaction.

Active learning is an 'instructional method that engages students in the learning process' and uses activities introduced in the classroom (Prince, 2004), which leads to students actively engaging and thinking in class, constructing knowledge and drawing conclusions and working together to understand problems (Farrell, Moog, & Spencer, 1999). The teacher is now a facilitator of knowledge, assisting groups and not answering questions directly (Farrell et al., 1999) and coaches rather than content experts (Warfa, Roehrig, Schneider, & Nyachwaya, 2014). The basic structure includes no lectures, students working in groups with assigned roles through a guided inquiry learning activity, and students take examinations individually at the conclusion (Farrell et al., 1999).

Problem-based learning, collaborative learning, cooperative learning and Process Orientated Guided Inquiry Learning (POGIL) are some examples of the types of active learning being used in education institutions (Prince, 2004). They require active student engagement and social interaction to understand and retain content (Moog & Spencer, 2008). These styles actively engage students in the classroom, which results in students experiencing improved learning (Farrell et al., 1999).

POGIL draws upon the learning cycle and constructivism to implement active student engagement in the session through the specifically scaffolded activity (Moog & Spencer, 2008). Students work in groups where each student has a self-appointed designated role that changes every session (Farrell et al., 1999). Roles include recorder, manager, presenter, technician and reflector, which require different skills to fulfil them (Farrell et al., 1999). By continuously changing the roles, students are exposed to a different set of skills that they can develop. It is also harder for students to become disengaged from the session as each person is vital to keep on task and complete the activity (Farrell et al., 1999).

In their groups, students complete a guided learning activity that covers the relevant content. The activity usually presents some data or a model followed by critical thinking questions (Farrell et al., 1999). The critical thinking questions require the student to make inferences, assumptions and conclusions to reveal fundamental relationships that are part of the unit's content (Farrell et al., 1999). Follow-up application questions may be included and look similar to homework questions where students apply the principles they have discovered (Farrell et al., 1999). In this process, it is important for facilitators not to answer questions directly as this does not promote engagement with the content (Farrell et al., 1999). Students work in their designated group to complete the guided learning activity, discuss questions and reach conclusions. All students have the opportunity to challenge and put forth their point of view, and it is up to the group to reach a consensus.

This approach to teaching has also resulted in a decrease in fail rates and withdrawal rates when it is implemented (Farrell et al., 1999). Compared to traditional lecture style units, students attain better results and positively embrace the guided inquiry learning for further studies (Farrell et al., 1999), indicating it is a better delivery model.

Communication

Communication is defined as 'the activity or process of expressing ideas and feelings or of giving people information' in the Oxford Advanced English Dictionary (Oxford Dictionaries, 2015). Communication is highly relevant in the learning environment as students are constantly expressing ideas to one another. One aspect of communication in the science classroom is critical thinking that is regarded as the evaluation of the worth, authenticity or accuracy of various propositions (Zeidler, Lederman, & Taylor, 1992). Along with the efforts of clear communication, unclear and imprecise communication can be expressed in the form of fallacies, incomplete arguments of normative reasoning that affect learning (Zeidler et al., 1992). Students formulating arguments and logical reasoning skills are required for communication and are shown not to be innate and require additional instruction (Zeidler et al., 1992).

Threshold Learning Outcomes

In 2011, the Australian Government introduced the Tertiary Education Quality and Standards Agency Act in which higher education standards framework, threshold learning outcomes (TLOs), were developed that described the minimum achievement an average student would graduate with by the end of their degree at an Australian university. This leads to the learning and teaching academic standards project where the threshold learning outcomes were defined for each discipline (Jones, Yates, & Kelder, 2011). Mapping into how each university teaches and assesses the chemistry threshold learning outcomes is being conducted (Schultz, Mitchell, & O'Brien, 2013).

Threshold learning outcome 4 is 'to be effective communicators of science' (Jones et al., 2011); however, this is not assessed frequently throughout a chemistry degree (Schultz et al., 2013). Communication is a broad topic and can include many elements of confusion due to its multiple forms (Colthorpe, Rowland, & Leach, 2013). Inquiry into the application of threshold learning outcome 4 needs to be conducted to ensure students are graduating with this achievement.

Productive Discourse

Productive discourse is essential to learning science (Duschl & Osborne, 2002), especially in the development of critical thinking (Zeidler et al., 1992). Developing students' discourse has been the focus of recent research, particularly in the learning sciences (Sawyer, 2014). Traditional teacher–student discourse involved an initiate-response-feedback system (IRF) and has been the most frequent discourse patterns in classrooms (McNeill & Pimentel, 2010). However, students lack the opportunity to work collaboratively and building on each other's ideas when using this discourse pattern (Tong, Chan, & van Aalst, 2018). Consequently, there has been growing interest in dialogical teaching in classroom discourse.

As students are completing group work, it is very important that the students communicate clearly and concisely to one another. The students must discuss questions in detail and arrive at a unanimous answer. The process of negotiating an answer includes individual students putting forth their view, challenging another's view and persuading others to their view (Gillies, Ashman, & Terwel, 2008). New ideas generated by exposure to teachers also expand new knowledge for the student (Gillies et al., 2008). This delicate process requires students to use established communication skills that are not always taught or developed correctly in class (Gillies et al., 2008). The tools required for facilitators to assist in this process are quite unclear, and its effectiveness is not known (Cole et al., 2014).

As the focus is not on teacher-directed learning, the way that a teacher interacts with the class dramatically changes to the teacher asking more questions and informal communication (Gillies et al., 2008). However, what the teacher communicates to the class is very important, and they must use prompts that promote student learning and class dynamics (Cole et al., 2014). To have effective active learning, the teacher is now a facilitator of knowledge and does not dictate answers to students (Farrell et al., 1999; Chen, Clarke, & Resnick, 2014; Hennessy et al., 2016). Instead, teachers guide students to the correct answer by using practical and conversational actions for students to gather knowledge (Lidar, Lundqvist, & Östman, 2006). Some actions teachers use are confirming moves, reconstructing, instructional, generative and reorientating to bridge the gaps in students' knowledge (Lidar et al., 2006; Warfa et al., 2014). The dialogic exchange between teachers and students is very important and effective in the active learning environment for discussions about the content (Warfa et al., 2014). The teachers provide most of the discourse to what the students are learning (Duschl & Osborne, 2002). For students to reach an understanding of the content, the students need to work together to build knowledge using the collective groups and facilitators that can be a hurdle (Cole et al., 2014). Some other facilitation cycles can be seen in the facilitator–student interaction of elicitation–student response–teacher elaboration (ERE) followed by proposition–discussion (PD) which is then repeated until the material is exhausted (Bowers & Nickerson, 2001). The

ERE and PD cycle is an effective dialogic exchange between teachers and students in collaborative learning.

Despite a large amount of analysis into the specific communication between teachers and students, norms will develop in class (Bowers & Nickerson, 2001; Gillies et al., 2008; Warfa et al., 2014). These norms will influence the students' engagement and practices in class. The students are also limited by their knowledge of the discourse they are learning, and when learning science, this limits the extent in which students can engage in dialogic argumentation, construction of explanation and evaluation of evidence (Duschl & Osborne, 2002).

Communication is an important skill and employers look for similar to teamwork skills. In a study conducted by Prince (2004), engineering students received direct instruction about teamwork skills when completing collaborative learning and successfully developed teamwork and social skills in the engineering students. The investigation into whether communication skills can be transferred in a similar manner is, therefore, important.

Talk Science Primer

To facilitate the process and acquisition of appropriate communication strategies students can use in class, the Talk Science Primer was developed to guide teachers in implementing academically productive talk in classrooms (Michaels & O'Connor, 2012). The Talk Science Primer was developed from American research, funded by the National Science Foundation, in primary and secondary school classrooms from years K-12. It is built upon previous research on promoting academically productive, accountable talk in the classroom (Chapin, O'Connor, & Anderson, 2009; Resnick, Michaels, & O'Connor, 2011). The need arises to instruct students and teachers on how to 'talk' because not all talk is productive or able to promote deeper conceptual understanding from students. There needs to be established beliefs, rules, purposes and formats to establish academically productive talk that is possible in the classroom. Discussions are characterised into four different types: elicitation, consolidation, data and explanation discussions that are all applicable to the classroom environment (Michaels & O'Connor, 2012).

Talk formats are classroom structures that are conducive to students having engaged discussions, which may be in the form of whole group, small group work or partner talk time. In addition, talk moves are strategic teacher-directed instruction to facilitate discussions in class (Michaels & O'Connor, 2012). The talk moves use a four-step goal sequence that students move through to develop communication skills for discussion. The students move through each goal and acquire new communication strategies at each stage. Each stage requires students to engage more in the discussions to achieve the next goal and continue until all goals are mastered. The Talk Science Goals and key outcomes are illustrated in Fig. 14.1.

The talk moves are most frequently in the form of probing questions but also include pauses and think time for students (Michaels & O'Connor, 2012). The outline

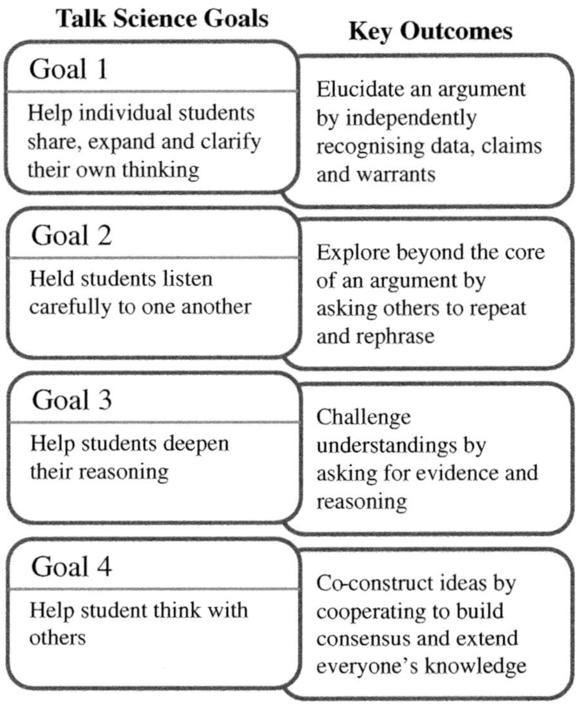

Fig. 14.1 Talk Science Goals illustrate how the pedagogical intervention will give rise to key outcomes that promote generative dialogic student discourse (adapted from Chapin et al., 2009)

detailed in Talk Science Primer assists teachers to create a culture in which students can clearly communicate to aid discussions about important concepts in class.

Context

Curtin University has adapted active learning using productive discourse for first-year chemistry classes replacing bulk lectures in which the content was delivered. The signature pedagogy in our first-year chemistry units is intended to foster these higher-order skills. The four first-year chemistry units have active learning principles implemented. Students attend a 2-h workshop each week where they work in groups of three to complete a guided inquiry learning activity. When students enter the workshop for the first time, they are randomly placed in groups of 3 and complete the first guided inquiry learning activity. Each student works with the same group throughout the semester where possible. Each group member takes on a role each week, and the roles change every week. Throughout the workshop, report outs are conducted and the class discusses certain key questions. At the end of the workshop, the group completes a short assessment that covers the main topics of the workshop session. The assessment is marked, graded and returned to the students the following week. The outcomes of the workshops at Curtin can be simplified into three separate outcomes: (1) each student has learnt the material covered, (2) each student has ensured that all members of their team have learned the material, and (3) a group assessment paper is submitted by each team. Since changing the teaching style, students' average final marks have increased between 6 and 15% for all units. The failure rates, which were well over 20% with the traditional lectures, dropped well below 20% in the new teaching style. Students were engaged in their learning and supported by feedback from their peers and facilitators. The students also attend a weekly one-hour lecture and a fortnightly 3-h laboratory.

Little is done to explicitly develop or support productive discourse, and nothing is known of its efficacy. There is no direct instruction provided to the students about working in groups or communicating effectively in groups. An assumption is made that the students can communicate effectively in their groups innately. However, this may not always be the case. Often, groups can struggle to work together and communicate effectively. The lack of group communication can negatively affect student work and engagement in the unit leading to poor results.

Aims

This project aims to develop instructional strategies that promote dialogic student discourse, as well as investigate and analyse how students communicate in the first-year chemistry workshops while completing the IC unit at Curtin University. The primary activity of this project is to adapt an instructional strategy called the Talk Science Primer (Michaels & O'Connor, 2012) used in K-12 classrooms with appropriate and contextualised learning objects to scaffold and develop productive discourse between students. Specific research questions include:

- Can the Talk Science Primer, which was developed for use in primary and secondary school, be adapted for use in first-year chemistry to promote generative dialogic student discourse?
- What are students' perceptions of interventions developed to help enhance group communication skills?

The purpose of this chapter is to describe the development of the interventions, adapted from the Talk Science Primer, and provide an insight students' perceptions of group communication.

Methodology

This study used a mixed methods approach (Towns, 2009) in a quasi-experimental sequential design (Ivankova, Creswell, & Stick, 2006). Both quantitative and qualitative data are widely used in educational research (Bhattacharyya & Bodner, 2005;

Bodner, MacIsaac, & White, 1999; Coe, McDougall, & McKeown, 1999). Qualitative and quantitative data were collected from information obtained from student responses for the interventions.

Interventions

In order to determine the efficacy of the interventions, each semester the cohort was split into two groups, intervention and non-intervention groups. The intervention group received direct instruction on developing communication strategies, while the non-intervention group did not. Interventions were implemented in the first 10 min of the workshop for the relevant weeks that related to bonding content. The students completed the interventions in their usual groups they worked in. The interventions were self-contained and required no additional work to be done by the students before or after the intervention.

In the week following the specific intervention, a summary of the activity was published in the learning management system (LMS), Blackboard, for students to review. The material published on LMS included question guides and prompts students could print out to take to the following workshops.

Participants

The IC cohort was selected to be involved in the research because the participants had little to no prior knowledge of chemistry. The students also came from a diverse background and a variety of degrees such as health science, nutrition and physical sciences. The students were identified as requiring additional support as they were new to learning the language and discourse of chemistry, which made them excellent candidates to investigate communication.

Cohort A consisted of students from semester 1. In total, three intervention workshops had 236 students and three non-intervention workshops had 179 students. The first versions of the interventions were trialled with Cohort A. Cohort B consisted of students from semester 2. In total, two intervention workshops had 103 students and two non-intervention workshops had 86 students.

Results and Discussion

Design and Implementation of Interventions

Theoretical Grounding

Each of the interventions was designed using the instructional strategy, Talk Science Primer, which was developed for use in K-12 classrooms to develop productive discourse in classrooms (Michaels & O'Connor, 2012). The Talk Science Primer was developed from American research, funded by the National Science Foundation, in primary and secondary school classrooms from years K-12. It builds upon previous research on promoting academically productive, accountable talk in the classroom (Chapin et al., 2009; Resnick et al., 2011). These interventions were completed in the groups students normally work in during the workshop. Communication in their normal group is very important in the workshops. Material on the LMS was selectively released so only students in the intervention group could access the information.

The first intervention targets *Goal 1: help student share, expand and clarify their own thoughts* by using 'probing questions'. The intervention was designed to initiate discussion about common issues when students work in groups as students come from a variety of backgrounds and have different experiences. The objective of the intervention was to have students discuss common issues and potential solutions for the issues. The intervention concludes with 'probing questions' as a strategy of how Goal 1 can be reached.

In the second intervention, students completed an activity on *Goal 2: help students listen carefully to one another* by establishing 'group norms'. The activity was designed to clarify each student's expectations of how the group should work together. Part of the activity covered expectations that could be considered normal etiquette such as not interrupting each other, coming prepared, notify someone if you are not attending or treat each other with respect and no judgement. This activity was designed to talk about these ideas and for the group to decide which ones were important. The objective of the activity was for groups to discuss group practices and decide on 'group norms' (rules) the group would agree to abide by to work together throughout the semester.

The third intervention incorporated both *Goal 3: help students deepen their own reasoning* and *Goal 4: help student engage with others reasoning* to develop argumentation strategies. The activity was designed to teach students to engage in argumentation that would promote discussions in workshops. Students were able to practise argumentation in the workshops with their group members in a constructive manner. The objective of the intervention was to identify the purpose and use of argumentation to support reasoning and engagement in discussions.

Intervention 1

Intervention 1 was conducted in week 2. Groups were asked to discuss and answer the following three questions:

- What do you think is the number one barrier for successful group work?
- What is your group most concerned about when communicating to each other?
  - Misunderstanding each other
  - Unable to explain using proper terms
  - Clarification and expressing your thoughts
  - Cannot develop from previous ideas
  - Personality clashes.
- What are some of the approaches your group could do to avoid these?

Cohort B was asked to answer some additional questions:

- What do you think communication is?
- How is this hindered?
- How do you deal with this?

Groups were allowed a couple of minutes to discuss questions before submitting answers. Cohort A submitted answers via an online portal called *Socrative*, while Cohort B did not due to time constraints. The answers were then discussed as a class with students volunteering their responses to the class.

In the final task, groups wrote three 'probing questions'. The purpose of probing questions was to elicit further explanation and clarification of student responses. The question was then designed to 'probe' into the reasoning to promote discussions. The Talk Science Primer includes teacher prompts (probing and redirecting questions) to guide students in the communication goals. As the teachers' role in the workshop includes less student contact time, the student themselves needs to take on the teacher thinking and uses the questions to achieve the goals. Therefore, the three probing questions come from the list of teacher prompts.

Probing question examples (the list of teacher prompts) were displayed on the projector, so students could model. They could directly copy or come up with some of their own in the same style. The students copied three questions on a sheet of paper and left the questions visible during the remainder of the workshop.

At the end of the workshop, the groups logged back onto *Socrative* to complete two more questions. The class did not discuss the questions collectively; instead, the questions were used as feedback. The questions were:

(1) Which of the questions your group came up with at the start of the day was used most frequently?

(2) Were the questions used in your discussions today?

   (a) No, I did not understand what to do

   (b) No, they were not used

(c) Yes, they were used
(d) No, they were not used but might be in the future.

In the following week, a full list of 13 probing questions was published on the LMS for use for the rest of the semester (see Appendix 1).

Intervention 2

Intervention 2 was conducted in week 4 in semester 1 and week 3 in semester 2. Cohort A used an online brainstorming site, called *GroupMap*, to brainstorm good and bad group practices. These ideas were listed in the first column called 'Aspect'. Students were also asked to list group rules in the second column called 'group norm' to match the good/bad group practices listed in 'aspect'. The group rules were discussed to ensure the groups would implement the good group practices and avoid the bad group practices.

Cohort B completed a worksheet instead of using GroupMap after redesigning Intervention 2 (see Appendix 2). The worksheet was distributed one per group for students to complete. The redesigned Intervention 2 was framed within the context of conflict resolution as conflict management was identified as an area that groups needed assistance. Conflicts occurred in groups and students found it difficult to manage them to reach an acceptable resolution. When Intervention 2 was delivered to students, it was highlighted as a key reason to establish the group norms. It was conceptualised to the students that if conflicts occurred later on in the semester, the student could then refer to the rules that they had agreed to abide by in this intervention to resolve the conflict.

In the following week, a summary sheet was produced by collating all the students' responses from the different workshops and published on the LMS. The students could see other group ideas and adopt them for use during the rest of the semester.

Intervention 3

Cohort A completed Intervention 3 online through LMS over the tuition free week (between weeks 4 and 5) and had an in-class summary in week 5. The students had a choice to either read information about the argumentation model (see Appendix 3) or watch a video. The students then answered a few short questions on the model to reinforce the idea of the model (see Appendix 4).

Toulmin's model of argumentation was used which identifies components of an argument (Toulmin, 1958; Cole, Becker, Towns, Sweeney, Wawro, & Rasmussen, 2011; Hitchcock & Verheij, 2006). The short video explained what an argument was, the purpose of an argument in a scientific context and the different parts of Toulmin's model that contribute to a strong argument as shown in Fig. 14.2. The video was 87 s long and then directed students to the worksheet.

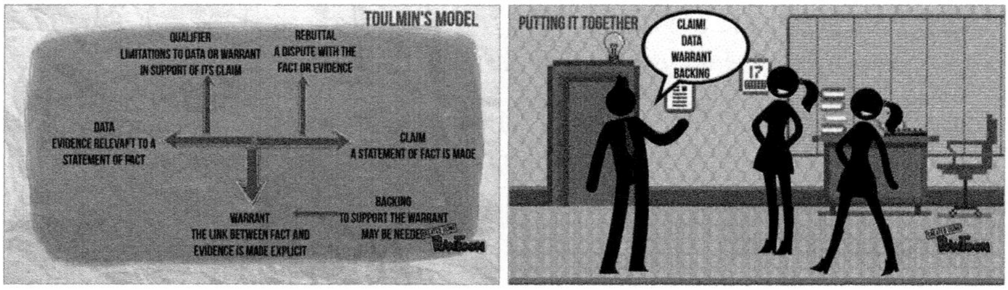

Fig. 14.2 Intervention 3 video (screenshot) describing argumentation used with Cohort A

The worksheet was two pages long. The first page contained information that the video provided and covered the same concepts, just presented differently. The information also provided a written argument example with all the components labelled. The second page was designed to be completed by all the students to check the students' understanding of Toulmin's model (Cole et al., 2011; Hitchcock & Verheij, 2006). It included six short questions that the student could complete identifying what they could use in their group (see Appendix 4). The last question specifically asked the student to describe how it could help them in their workshop.

The argumentation activity was presented in the workshop in week 5 (see Fig. 14.3). The information presented was directly covered in the video and information sheet. There was no class discussion allocated, and most students had not completed the online activity when asked. Therefore, this was the first time many had seen the argument and learnt about it.

Intervention 3 was completely redesigned for Cohort B. The intervention covered Goals 3 and 4 and was changed to an in-class activity. A worksheet was developed (see Appendix 5) that let students practise argumentation and exploration of their own and others' reasoning with their group in the workshop. The worksheet developed was linked to chemistry concepts as this provided practice on how it could be actively used in the workshops and provided an extra reward for participating in the intervention.

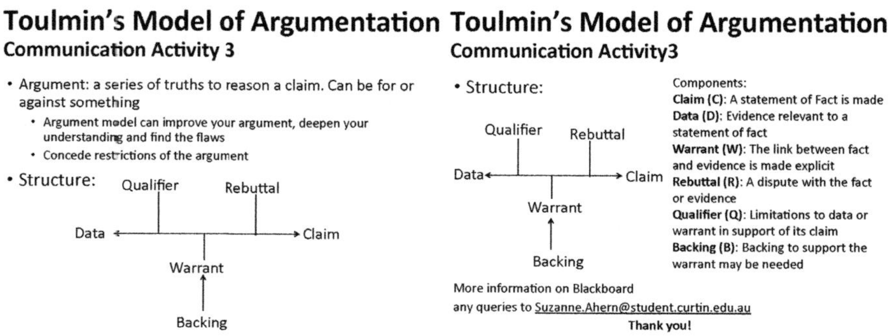

Fig. 14.3 Argumentation information presented to the students in Intervention 3 of Cohort A

First, it was explained what an argument was (series of truths to reason a claim and not a yelling match) and the purpose of engaging in argumentation (deepen intellectual understanding, find flaws in reasoning and improve your argument). Each group member then individually chose two basic structural terms out of a proton, neutron, electron, atom, nucleus and ion that they thought made the best pair for a particular reason. Each group member then took turns to argue and persuade the other group members why their pair was the best combination. The students were also provided a list of questions they could use to engage in argumentation and discussion (see Appendix 6). The list of questions included the teacher prompts from the Talk Science Primer. All the worksheets were collected at the end of the workshop. A summary sheet was created that collated all the different student responses to Intervention 3 and published on the LMS.

Intervention 4

An additional intervention was created for Cohort B. Intervention 4 was designed to summarise and briefly refresh the previous interventions' objectives, goals and the communication activities. Intervention 4 was designed to be completed as a group in a different way than previous interventions. As this would likely include many questions since *Socrative* was successful, *Socrative* was used in this intervention. The objective of *Socrative* was to have groups revisit the major communication goals and themes that were discussed in the intervention. The quiz was set up with a mixture of 9 short answer, multiple-choice and true/false questions (see Appendix 7). Of the 9 questions, 4 related to Interventions 1, 2 related to Interventions 2 and 3 related to Intervention 3.

Student Response to the Interventions and Areas of Improvement

Intervention 1

Student answers are shown in Fig. 14.4. The majority of groups (77.5%) identified communication as the number one barrier for successful group work (Question 1). The identification of communication is significant as it indicates that students already identify communication as an important factor when doing group work. Other issues raised were participation (9.0%) and co-operation (4.5%). When groups are communicating, the majority (40.4%) of groups selected clarification and expressing your thoughts as the main concern (Question 2). Misunderstanding each other was also a concern most groups had (29.8). Students answered with a variety of approaches groups could use to avoid common problems (Question 3). The top three responses were using explanations (20.7), include everyone (13.5%) and communicate more

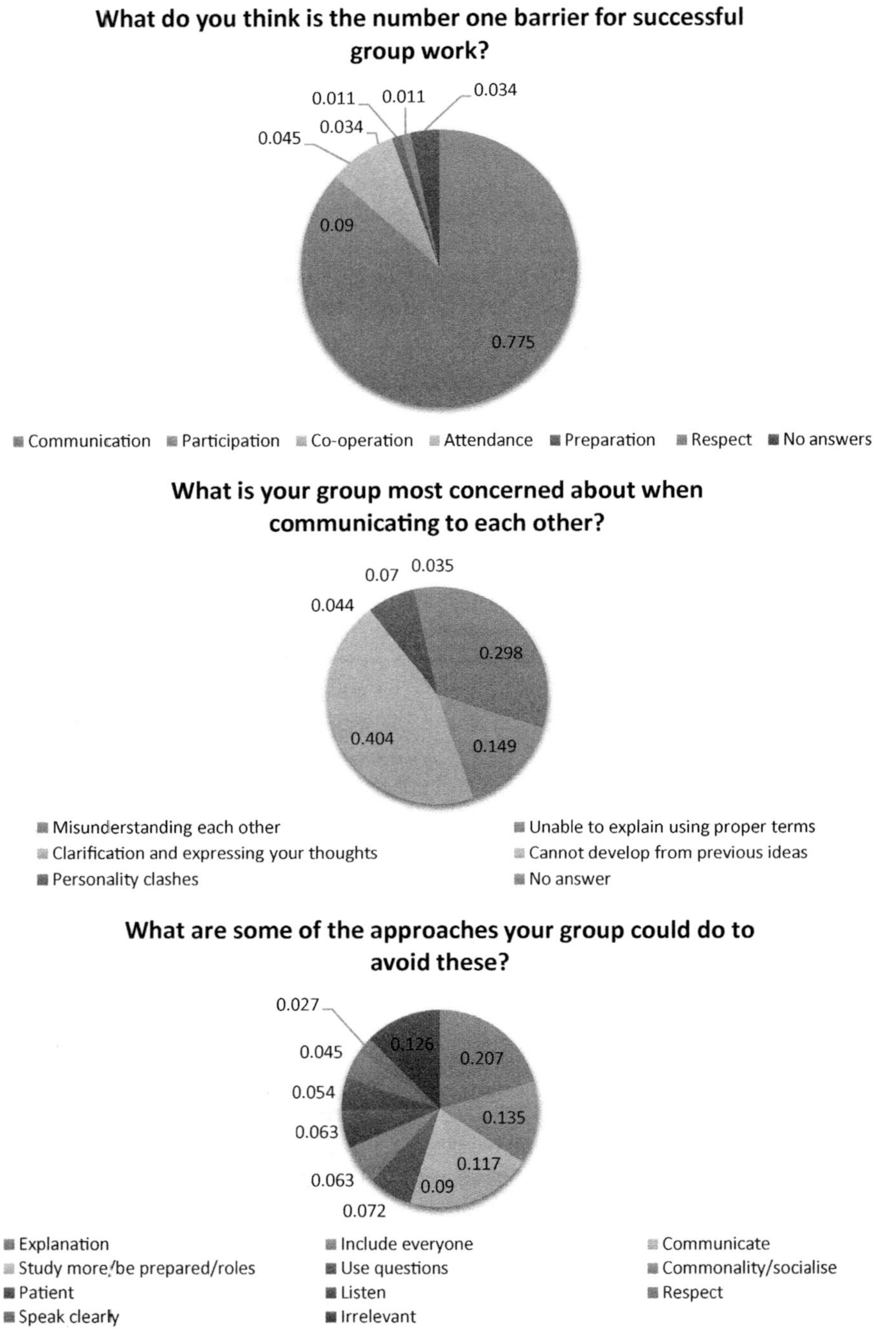

Fig. 14.4 Student responses to Intervention 1

(11.7%). However, the responses were vague and did not show a clear idea of how to overcome the issues indicating that students were not sure themselves.

Overall, Intervention 1 ran smoothly and students participated in it. There were no major obstacles or hesitations from the majority of students. Discussion of group communication flowed freely in the groups and the workshop.

Intervention 2

Initially, when Intervention 2 was explained to Cohort A, students struggled to understand the objective of the intervention. Students struggled with conceptualising the purpose of this intervention and misunderstood the term 'group norms'. The students found it difficult to link 'group norms' to their group's practice. Therefore, a lot of time was used explaining what to do in the groups that did not allow enough time for brainstorming and discussion. Groups eventually came up with ideas, and an example of one workshop's brainstorm is shown in Fig. 14.5.

Some technical difficulties were experienced with using *GroupMap*. Therefore, Intervention 2 was significantly redesigned and improved for subsequent use with Cohort B. The technical difficulties and unfamiliarity of *GroupMap* were avoided by using a worksheet instead. The worksheet allowed the intervention to keep within the allotted time and promoted student discussion. There was significantly more participation for Cohort B compared with Cohort A. Almost all of the groups dis-

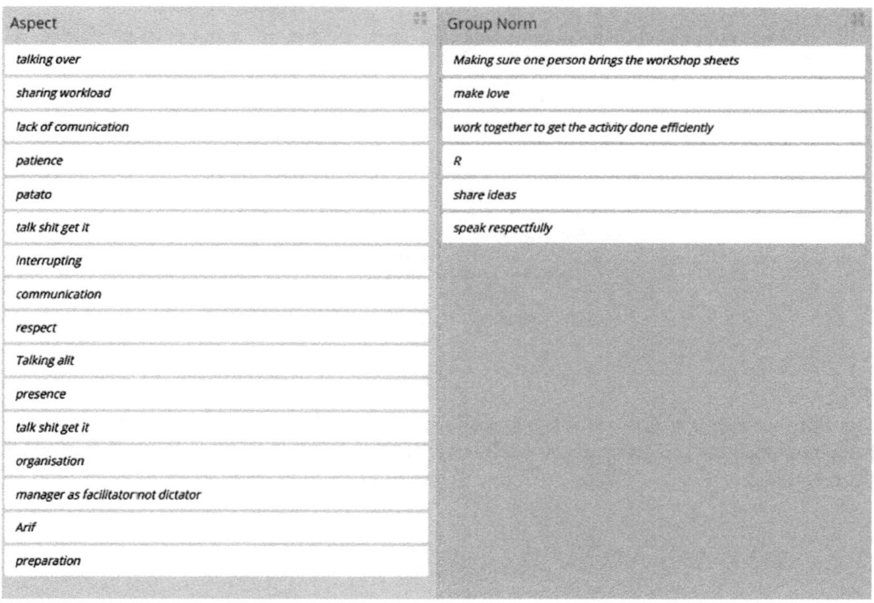

Fig. 14.5 Example *GroupMap* of a workshop's brainstorm in Intervention 2 from Cohort A

cussed Intervention 2 in class and were able to produce ideas. Figure 14.6a–c shows combined ideas for Cohort A and Cohort B and the percentage of responses of the listed ideas. The top five good group practices include: equal input from all group members, listening to each other and respect, communication and active participation and attend. The top five bad group practices include: talking over people, not participating, narrow mindedness, being disrespectful and not attending. The top five suggested group norms include: equal contribution, all must participate, listening, respect and come prepared.

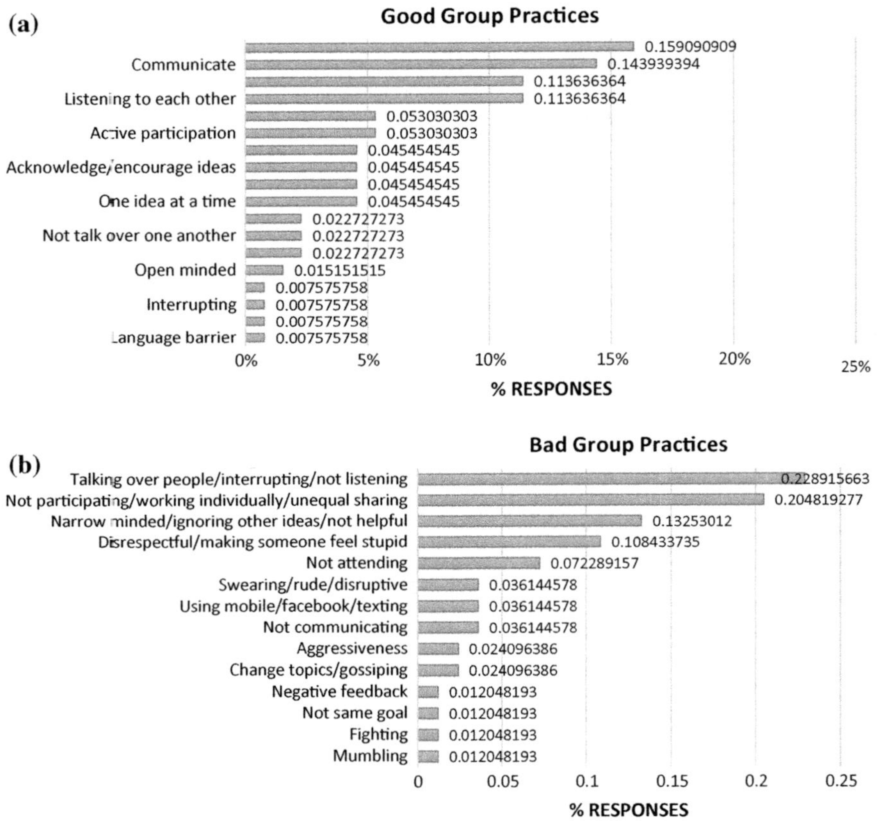

Fig. 14.6 **a** Student responses for good group practices to Intervention 2. **b** Student responses for bad group practices and group norms for Intervention 2. **c** Student responses for group norms for Intervention 2

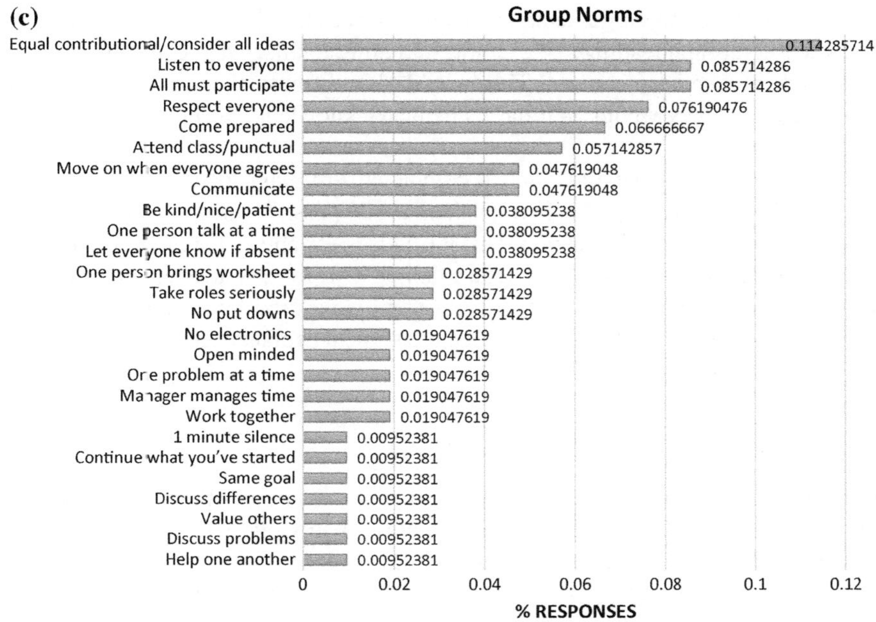

Fig. 14.6 (continued)

Intervention 3

The intervention group of Cohort A completed an online activity for Intervention 3. It was therefore up to the motivation of the student to complete the intervention outside any normal class times. All previous interventions were completed in the first 10 min of the workshop. This promoted student participation despite the fact that they were all voluntary. As Intervention 3 was not conducted during normal class times, completion rates were lower. Only about half of the students from the intervention group completed Intervention 3 in Cohort A. The overall success of Intervention 3 was diminished due to the lack of participation.

Another inherent hurdle of the activity was that the students were completing it individually. For other interventions, students discussed and applied the concepts to their groups. This made the activities quite successful as it promoted discussions about group work that was directly relevant for the students at that point in time as they were with their group. However, the online individual activity meant group discussion about argumentation was not possible as the students were not in their groups. As the purpose of this intervention was to develop group communication, it was obvious that it needed to be a group activity similar to the previous interventions and this was changed for subsequent use.

A further hurdle of the argumentation activity is that it is a difficult concept to understand that was presented with no direct relevance to the chemistry students.

As the concept of the argumentation model is not straight forward, it cannot be easily learnt through reading and examples. In order to improve the model, it should be linked to chemistry content. This would also increase the reward for student participation as they would gain more chemistry knowledge if they participated in the intervention.

The intervention group of Cohort B completed a redesigned version of Intervention 3, as described on Page 12. In total, there are 30 possible pairs although not all of the pairs can be chemically related. This introduces variability in responses, and this initiated discussion from students. One example of a group's answer is represented in Appendix 5. However, only 8 different pairs were used by students and this is summarised in Table 14.1 with the relevant chemical explanation.

Intervention 4

Students demonstrated good understanding and recollection of the main ideas from the interventions while completing Intervention 4 in Cohort B (see Table 14.2). Students were able to recall the goals for each intervention, the techniques that were taught in the interventions, and identify examples of the techniques. Students were also able to provide examples that they were using in their groups from the interventions.

Socrative was used for Intervention 4 that was a different format to the other interventions. The different format of multiple choice questions was necessary as it made Intervention 4 different and unique. Part of the success was due to the different format, and keeping it different is essential to reaffirm the ideas of the previous interventions. However, the questions asked require further development to match the allocated timing of Intervention 4 while also using the online platform Socrative.

Conclusions and Implications for Practice

This project aimed to develop instructional strategies that promote dialogic student discourse, as well as investigate and analyse how students communicate in the first-year chemistry workshops while completing the IC unit at Curtin University. The primary activity of this project is to adapt an instructional strategy called the Talk Science Primer used in K-12 classrooms with appropriate and contextualised learning objects to scaffold and develop productive discourse between students. This book chapter focused on two main areas—the design of interventions and students' perceptions of group communication. The specific research questions for this study can now be answered.

Can the Talk Science Primer, which was developed for use in primary and secondary school, be adapted for use in first-year chemistry to promote generative dialogic student discourse?

Table 14.1 All pairs reported during Intervention 3 (total 28 groups)

Pair and argument		No. of responses
1	Pair: electron, proton	14
	Argument: equal charge value but just opposite sign. Both contribute to overall charge and influence the behaviour of atoms	
2	Pair: neutron, proton	13
	Argument: both are in the nucleus, and you cannot have one subatomic particle without the other except for hydrogen (i.e. 10 protons would not exist without neutrons). They are the most dense part of an atom	
3	Pair: electron, ion	10
	Argument: an ion is defined by the number of electrons it has (normal amount or more or less). An ion is created by either losing or gaining electrons. They are important when forming bonds	
4	Pair: atom, ion	6
	Argument: an atom has the capability to be turned into an ion by losing or gaining electrons	
5	Pair: neutron, nucleus	5
	Argument: it gives mass to an atom which is a key component	
6	Pair: proton, ion	1
	Argument: the charge of an ion is calculated by the amount of electrons (given or taken) relative to the protons	
7	Pair: atom, nucleus	7
	Argument: you cannot have one without the other. Whenever there is an atom, there is always a nucleus. The nucleus contains the atomic weight of an atom	
8	Pair: proton, atom	2
	Argument: if you change the number of protons in the nucleus, the atom (or element) changes and can have different characteristics. This change of protons corresponds to different atomic number	
9	Pair: neutron, atom	1
	Argument: neutrons are part of atoms	

Table 14.2 Summary of key points students answered in Intervention 4

Activity 1: Probing Questions	Activity 2: Group Norms	Activity 3: Argumentation
• Accurately identify Goal 1 • Listed some concerns about communication, team work and time management • Most identified probing question technique • Could identify probing question • Some common questions students used were what do you think? Can you explain that?	• Identify Goal 2 • Identified group norm examples • Provided varied group norms group used such as everyone to come prepared, manager to keep team on track, be respectful	• Identified Goals 3 and 4 • Identified the purpose of engaging in argumentation • Identified appropriate questions to use

Materials were developed in the form of group activities to target communication using the Talk Science Primer Talk Goals. Four interventions were developed after an iteration and refinement. Intervention 1 was very successful, by achieving what it was designed to achieve, and used an online quiz portal, *Socrative*, to instruct students on how to use 'probing' questions to elicit, share and clarify ideas. Students had difficulty in conceptualising the first version of Intervention 2 and Intervention 3, which was conducted online, had a low participation rate. The ideas presented in these interventions were also abstract, and students found it difficult to relate the intervention's objective to their chemistry studies. Therefore, interventions were redesigned for subsequent use. The first three interventions were completed as group discussions with the use of a worksheet and the final intervention used *Socrative*. All interventions were conducted in class, and all students were engaged. The level of the activities was appropriate for tertiary students. This demonstrates that in-class activities are more effective so any future interventions should be all done in class.

What are students' perceptions of interventions developed to help enhance group communication skills?

Students were able to identify communication as the number one barrier for successful group work. Other issues raised were participation and co-operation. When groups are communicating, the majority of groups believe 'expressing your thoughts' and 'misunderstanding each other' as main concerns. Students had a variety of approaches groups could use to avoid common problems. The top three responses were using explanations, include everyone and communicate more. Some group practice examples include sharing workload, listening to each other, respect, communication, active participation and organisation. Bad group practices include talking over people, not participating, narrow mindedness, being disrespectful and not attending. Some key group norms identified by students were to speak respectfully, assign tasks such as one person to bring the worksheet, everyone must participate, listening, respect and come prepared. This demonstrates that the interventions were able to illicit student perceptions of communication that can use for group work in the future. Simply getting students to think about group work and what expectations they have, makes them more aware of the importance of communication and may influence them using the suggested prompts in future group situations.

Acknowledgements We would like to thank the IC unit coordinator, Lise Fouché, who allowed us to research her unit, all the workshop facilitators who supported us running the interventions in their classes and the students who participated in our study. This study was approved by the Curtin University Human Research Ethics Committee (Project Number SCI-54-14).

Appendix 1: List of Probing Questions Released to Students After Completion of Intervention 1

Probing Questions

CHEM1003 Introduction to Chemistry

During the workshop session in week 2, your groups discussed problems when completing group work and ways to avoid them. One solution was for you to ask questions within your group to probe how much each group member understood the material. Groups in your workshop came up with 2 or 3 questions and had them on the desk to refer to during the session if needed. Here is a more detailed list of questions for you to print and take with you to your workshop to use in your group discussions.

What do you mean by [term]?

Can you explain [term]?

Can you explain using another example?

Can you expand on ……?

Do you mean …… by ……?

When you said [phrase] are you referring to ….?

What do you think [other group member]?

How sure are you?

What are you unclear about?

What is the reasoning behind that?

How are ….. and …… connected?

What is the question asking?

What do you know already?

Appendix 2: Example Group Worksheet for Intervention 2

Group Norms

CHEM1003 Introduction to Chemistry

Group Norms are rules that your group agree to abide by in order for your group to function and be accountable. They are extremely useful for dealing with confrontations or if someone is not contributing. Some group norms were set up in activity 0 in week 1 of semester. Some group norms are implicit such as attending every workshop however it is always helpful to define these and the consequences (such as bringing lollies for everyone if you miss a class!). Today, your group is going to set up some group norms to assist your group communication.

Firstly, complete the following table by listing good group practices (such as not talking over each another) and bad group practices (such as interrupting each other).

Good Group Practice	Bad Group Practice
Listening to eachother	Not communicating
Asking questions to include	Talking over people
all group members.	Using your mobile
Working as a team,	Making someone feel stupid.
equal input.	

With all of the good and bad group practices you have brainstormed, you want to aim for your group to have all of the good practices and avoid the bad practices. To do this, your group should create some group norms (rules) to follow to promote the good practices and avoid the bad practices.

	Group Norms
1	All team members must participate. (Recorder, Presenter, Manager
2	Listen to your team members and respect their ideas
3	Do not talk over team members
4	Let your team members know if you will not be on time or if your unable to make a session.

Appendix 3: Information on Argumentation Using Toulmin's Model that Was Provided to Students Online for Intervention 3 for Cohort A

Argumentation

CHEM 1003 Introduction to Chemistry

During the workshops we have been working on improving your communication. Improved communication can allow you to reach deeper understanding of the content. This short activity will take you through a model of an argument. This can be applied to your discussions in your workshops.

An argument is a series of truths to reason a claim and it can be for or against something. Arguments are not fights as fights rarely include true arguments. An argument model can be used in your academic studies to improve your argument, find the flaws and deepen your intellectual understanding. In the model we will examine, Toulmin's model, we can also concede the restrictions on the argument.

Toulmin's model is based around the following relationships between components:

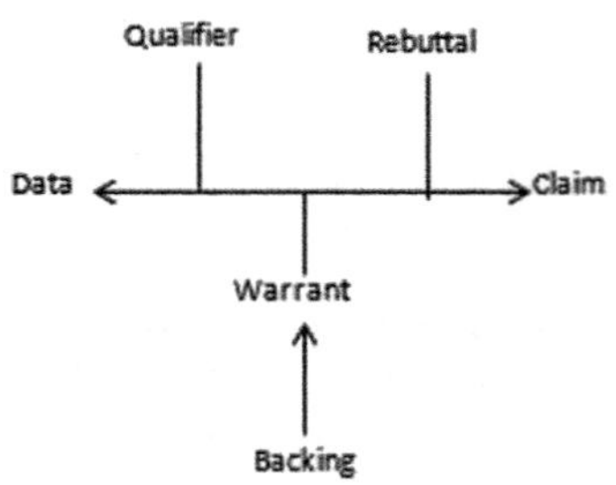

Claim (C): A statement of Fact is made

Data (D): Evidence relevant to a statement of fact

Warrant (W): The link between fact and evidence is made explicit

Rebuttal (R): A dispute with the fact or evidence

Qualifier (Q): Limitations to data or warrant in support of its claim

Backing (B): Backing to support the warrant may be needed

Below is an example of an argument with all the different components labelled by () after the sentence.

Smoking in public places should be banned (C) because it puts other people, especially children and pregnant women, at risk of breathing smoke from cigrettes (D). Smoking in public places also endangers people who have respiratory ailments (D). Recent studies show that almost 80% of those who ingest second-hand smoke from public smokers have a higher risk of getting respiratory problems than smokers themselves (D).

Banning an act that causes problems to innocent civilians is helpful in many ways (W). If smoking in public places is banned, we actually reduce or totally eradicate the danger of putting non-smokers at risk of developing ling and heart problems (B). Moreover, if we ban smoking in public places, we also stop the smokers from further increasing their chances of acquiring health problems from themselves (B).

While it can be said that not all people who smoke in public areas are always causing harm to others, it remains a fact the smoking per se is a cause of health problems (R). It is not enough to say that the size of affected people are relatively just a small fraction; plenty or few, one person at risk is more than enough (R). It is only the case that smoking in public places, therefore, should be banned (C).

Appendix 4: Argumentation Worksheet with Questions Relating to the Argumentation Model for Students in Cohort A to Complete

---

Argument Activity

1. Why are we applying the argument model? What uses does it have?

2. Define these components:

a) Claim

b) Warrant

c) Rebuttal

3. Label the following sentences which correspond to the argumentation components.

You should buy Bleach by Bleach and use it to wash your clothes (__). Studies show that clothes are 50% whiter after using our product (__). People want whiter clothes which is what our product will achieve (__). Whiter clothes have the advantage of looking cleaner (__) unless you want to look dirty (__). Therefore, you should use Bleach by Bleach to wash your clothes (__).

4. Using the above examples, can you comment on any common groupings of the components (i.e. what is the claim followed by?)

5. What is repeated at the end of an argument?

6. How could this help in your workshops?

Appendix 5: Example of a Groups Completed Intervention 3 Worksheet

---

Engaging in Argumentation

CHEM1003 Introduction to Chemistry

An argument is a series of truths to reason a claim and it can be for or against something. Arguments are not fights as fights rarely include true arguments. An argument model can be used in your academic studies to improve your argument, find the flaws in your reasoning and deepen your intellectual understanding.

Today, we are targeting Goal 3 and Goal 4 which is help students deepen their reasoning and help students engage with others reasoning respectively. We are going to do this by a short exercise.

Firstly, individually choose two of the six words that you think is the best pairing (don't tell your group members).

neutron	atom
proton	nucleus
electron	ion

Now, in your group, take turns at trying to persuade other group members as to which pairing is the best. Make sure you have a reasoned argument as to why and the other group members can ask you questions (see back for questions)!

	Arguments
1	Paired words: neutron + Nucleus
	It gives Mass to an Atom that is a key component
2	Paired words: Proton + electron
	Positive + negative charges
3	Paired words: Proton + ION
	the charge of the ion is shown by how Many electrons is given or taken.

Which was the best pair? 3

Why? Best explained

Appendix 6: List of Questions to Promote Discussion for Intervention 3

Example Questions to Use

Asking for evidence or reasoning

- Why do you think that?
- What's your evidence?
- How did you arrive at that conclusion?
- Is there anything in the text that made you think that?

Challenge or counterexample

- Does it always work that way?
- How does that idea square with this example?
- What if it had been _______ instead?

Agree/disagree and why?

- Do you agree/disagree?
- Are you saying the same thing as [group member] or something different? If it is different, how is it different?
- What do you think about what [person] said?
- Does anyone want to respond to that idea?

Add on

- Who can add on to that idea?
- Can anyone take that suggestion and push it a little further?

Explaining what someone else means

- Who can explain what [person] means when they said that?
- Who else thinks they could explain in their words why [person] came up with that?
- Why do you think they said that?

Appendix 7: Intervention 4 Socrative Questions

1. In week 2, we discussed some things your group was worried about when working together. What were some of the points your group discussed?

2. Techniques were discussed in week two to reduce the impact of group problems. What was one solution we explored in week 2?
   - (A) Coming prepared
   - (B) Asking probing questions
   - (C) Setting expectations
   - (D) Do nothing

3. What is an example probing question you could use in your workshops? Remember Goal 1 is to share, expand and clarify an individuals thoughts.
   - (A) You are wrong
   - (B) I disagree with that
   - (C) That makes sense
   - (D) Can you expand on that?

4. What was a common probing question you used in your groups during the workshops?

5. Goal 2 is to listen carefully to one another. What were some group norms (group rules) you could use as a group to achieve goal 2?
   - (A) Spend 1 minute per person to discuss an answer
   - (B) Everyone to come prepared and let other group members know if they are going to be absent
   - (C) Manager is responsible for making sure everyone is contributing equally
   - (D) Show respect to one another and not talk over anyone
   - (E) All of the above

6. What is a group norm (group rule) that assisted your group during the semester that you might recommend to another group?

7. Goal 3 is to deepen an individuals reasoning and Goal 4 is to engage with others' reasoning.What is the point of an argument?
   - (A) To yell at each other
   - (B) Deepen ones reasoning
   - (C) Find flaws in your reasoning
   - (D) Explain your point
   - (E) Answers B, C and D

8. Why was developing an argument important?
   - (A) To discuss the workshop questions in detail
   - (B) To yell at each other
   - (C) To persuade people to see your point of view
   - (D) No reason at all

9. What were some questions you could use to understand someones argument?
   - (A) Does it always work that way?
   - (B) Can you use another example?
   - (C) Ok, cool
   - (D) No it does not happen like that

References

Alexander, P. A. (2007). Bridging cognition and socioculturalism within conceptual change research: Unnecessary foray or unachievable feat? *Educational Psychologist, 42*(1), 67–73.

Bhattacharyya, G., & Bodner, G. M. (2005). "It gets me to the product": How students propose organic mechanisms. *Journal of Chemical Education, 82*(9), 1402–1407.

Bodner, G. M. (1986). Constructivism: A theory of knowledge. *Journal of Chemical Education, 63*(10), 873–878.

Bodner, G. M., MacIsaac, D., & White, S. (1999). Action research: Overcoming the sports mentality approach to assessment/evaluation. *University Chemistry Education, 3*(1), 31–36.

Bowers, J. S., & Nickerson, S. (2001). Identifying cyclic patterns of interaction to study individual and collective learning. *Mathematical Thinking and Learning, 3*(1), 1–28. https://doi.org/10.1207/S15327833MTL0301_01.

Brown, A. L. (1994). The advancement of learning. *Educational Researcher, 23*(8), 4–12.

Chapin, S. H., O'Connor, C., & Anderson, N. C. (2009). *Classroom discussions: Using math talk to help students learn, Grades K-6* (2nd ed.). Math Solutions.

Chen, G., Clarke, S. N., & Resnick, L. B. (2014). An analytic tool for supporting teachers' reflection on classroom talk. In *Learning and becoming in practice: The International Conference of the Learning Sciences* (Vol. 1, pp. 583–590).

Coe, E. M., McDougall, A. O., & McKeown, N. B. (1999). Is peer assisted learning of benefit to undergraduate chemists? *University Chemistry Education, 3*(2), 72–75.

Cole, R. S., Becker, N., Towns, M. H., Towns, M., Sweeney, G., Wawro, M., et al. (2011). Adapting a methodology from mathematics education research to chemistry education research: Documenting collective activity. *International Journal of Science and Mathematics Education, 10*(1), 193–211. https://doi.org/10.1007/s10763-011-9284-1.

Cole, R. S., Becker, N., & Stanford, C. (2014). Discourse analysis as a tool to examine teaching and learning in the classroom. In *Tools of chemistry education research* (Vol. 1, pp. 61–81). https://doi.org/10.1021/bk-2014-1166.ch004.

Colthorpe, K., Rowland, S., & Leach, J. (2013). Good practice guide (science) threshold learning outcome 4 communication. http://www.chemnet.edu.au/sites/default/files/files/GoodPracticeGuideScienceTLO4.pdf. Accessed December 2018.

Dewey, J. (1916). *Democracy and education: An introduction to the philosophy of education*. New York: Macmillan.

Driver, R., Asoko, H., Leach, J., Scott, P., & Mortimer, E. (1994). Constructing scientific knowledge in the classroom. *Educational Researcher, 23*(7), 5–12.

Duschl, R. A., & Osborne, J. (2002). Supporting and promoting argumentation discourse in science education. *Studies in Science Education, 38*(1), 39–72. https://doi.org/10.1080/03057260208560187.

Farrell, J. J., Moog, R. S., & Spencer, J. N. (1999). A guided inquiry general chemistry course. *Journal of Chemical Education, 76*(4), 570–574. https://doi.org/10.1021/ed076p570.

Gillies, R. M., Ashman, A. F., & Terwel, J. (Eds.). (2008). *The teacher's role in implementing cooperative learning in the classroom* (8th ed.). Boston, MA: Springer US. https://doi.org/10.1007/978-0-387-70892-8.

Hennessy, S., Drummond, S. R., Higham, R., Marquez, A. M., Maine, F., Rios, R. M., et al. (2016). Developing a coding scheme for analysing classroom dialogue across educational contexts. *Learning, Culture and Social Interaction, 9*, 16–44.

Hitchcock, D., & Verheij, B. (2006). Introduction. In D. Hitchcock & B. Verheij (Eds.), *Arguing on the Toulmin model* (pp. 1–23). Dordrecht: Springer Netherlands. https://doi.org/10.1007/978-1-4020-4938-5.

Ivankova, N. V., Creswell, J. W., & Stick, A. L. (2006). Using mixed-methods sequential explanatory design: From theory to practice. *Field Methods, 18*(1), 3–20. https://doi.org/10.1177/1525822X05282260.

John-Steiner, V., & Mahn, H. (1996). Sociocultural approaches to learning and development: A Vygotskian framework. *Educational Psychologist, 31*(3/4), 191–206.

Jones, S., Yates, B., & Kelder, J.-A. (2011, September). *Learning and teaching academic standards project science learning and teaching academic standards statement*.

Lidar, M., Lundqvist, E., & Östman, L. (2006). Teaching and learning in the science classroom the interplay between teachers' epistemological moves and students' practical epistemology. *Science Education, 90*(1), 148–163. https://doi.org/10.1002/sce.20092.

Mason, L. (2007). Introduction: Bridging the Cognitive and Sociocultural Approaches in Research on Conceptual Change: Is It Feasible? *Educational Psychologist, 42*(1), 1–7.

McNeill, K. L., & Pimentel, D. S. (2010). Scientific discourse in three urban classrooms: The role of the teacher in engaging high school students in argumentation. *Science Education, 94*(2), 203–229.

Michaels, S., & O'Connor, C. (2012). *Talk science primer*. Cambridge, MA: TERC. www.terc.edu. Accessed December 2018.

Moog, R. S., & Spencer, J. N. (2008). Chapter 1 POGIL: An overview. *ACS Symposium Series; American Chemical Society*, 1–13.

Oxford Dictionaries. (2015).

Palincsar, A. S. (1998). Social constructivist perspectives on teaching and learning. *Annual Review of Psychology, 49,* 345–375.

Phillips, D. C. (1995). The good, the bad and the ugly: The many faces of constructivism. *Educational Researcher, 24*(7), 5–12.

Piaget, J. (1952). *The origins of intelligence in children*. New York: International Universities Press.

Powell, K. C., & Kalina, C. J. (2009). Cognitive and social constructivism: Developing tools for an effective classroom. *Journal of Education, 130*(2), 241–250.

Prince, M. (2004). Does active learning work? A review of the research. *Journal of Engineering Education, 93*(3), 223–231.

Resnick, L. B., Michaels, S., & O'Connor, C. (2011). (Well-structured) talk builds the mind. In R. Sternberg & D. Preiss (Eds.), *From genes to context: New discoveries about learning from educational research and their applications* (pp. 163–194). New York: Springer.

Sawyer, R. K. (2014). *The Cambridge handbook of the learning sciences*. New York: Cambridge University Press.

Schultz, M., Mitchell, J., & O'Brien, G. (2013). Outcomes of the chemistry discipline network mapping exercises: Are the threshold learning outcomes met? *International Journal of Science and Mathematics Education, 21*(1), 81–91.

Spencer, J. N. (1999). New directions in teaching chemistry: A philosophical and pedagogical basis. *Journal of Chemical Education, 76*(4), 566. https://doi.org/10.1021/ed076p566.

Tong, Y., Chan, C. K. K., & van Aalst, J. C. W. (2018). Developing productive discourse among low-achievers in a knowledge building environment. In *Rethinking Learning in the Digital Age: Making the Learning Sciences Count. Proceedings of the 13th International Conference of the Learning Sciences (ICLS) 2018* (Vol. 2, pp. 879–886).

Toulmin, S. E. (1958). *The uses of argument*. Cambridge: Cambridge University Press.

Towns, M. H. (2009). Mixed methods designs in chemical education research. In D. M. Bunce & R. S. Cole (Eds.), *Nuts and bolts of chemical education research* (Vol. 976, pp. 135–148). Washington, DC: American Chemical Society. https://doi.org/10.1021/bk-2008-0976.ch009.

von Glasersfeld, E. (1993). Questions and answers about radical constructivism. In R. Tobin (Ed.), *The practice of constructivism in science education* (pp. 23–38). Hillsdale, NJ: Lawrence Erlbaum Associates.

Vygotsky, L. S. (1978). *Mind in society: The development of higher psychological processes*. Cambridge: Harvard University Press.

Warfa, A. R. M., Roehrig, G. H., Schneider, J. L., & Nyachwaya, J. (2014). Role of teacher-initiated discourses in students' development of representational fluency in chemistry: A case study. *Journal of Chemical Education, 91*(6), 784–792. https://doi.org/10.1021/ed4005547.

Zeidler, D. L., Lederman, N. G., & Taylor, S. C. (1992). Fallacies and student discourse: Conceptualizing the role of critical thinking in science education. *Science Education, 76*(4), 437–450. https://doi.org/10.1002/sce.3730760407.

Chapter 15
Improving the Assessment of Transferable Skills in Chemistry Through Evaluation of Current Practice

Madeleine Schultz, Glennys O'Brien, Siegbert Schmid, Gwendolyn A. Lawrie, Daniel C. Southam, Samuel J. Priest, Kieran F. Lim (林百君), Simon M. Pyke, Simon B. Bedford and Ian M. Jamie

Abstract The development and assessment of transferable skills acquired by students, such as communication and teamwork skills, within undergraduate degrees is being increasingly emphasised. Many instructors have designed and implemented assessment tasks with the aim to provide students with opportunities to acquire and demonstrate these skills. We have now applied our previously published tool to evaluate whether assessment tasks allow students to demonstrate achievement of

---

M. Schultz (✉) · K. F. Lim (林百君)
Deakin University, Geelong, VIC, Australia
e-mail: madeleine.schultz@deakin.edu.au

K. F. Lim (林百君)
e-mail: kieran.lim@deakin.edu.au

G. O'Brien
University of Wollongong, Wollongong, NSW, Australia
e-mail: gobrien@uow.edu.au

S. Schmid
The University of Sydney, Sydney, NSW, Australia
e-mail: siegbert.schmid@sydney.edu.au

G. A. Lawrie
University of Queensland, St. Lucia, QLD, Australia
e-mail: g.lawrie@uq.edu.au

D. C. Southam
Curtin University, Perth, WA, Australia
e-mail: D.Southam@curtin.edu.au

S. J. Priest
University of Adelaide College, Adelaide, SA, Australia
e-mail: samuel.priest@adelaide.edu.au

S. M. Pyke
University of Adelaide, Adelaide, SA, Australia
e-mail: simon.pyke@adelaide.edu.au

S. B. Bedford
Western Sydney University, Penrith, NSW, Australia
e-mail: s.bedford@westernsydney.edu.au

M. Schultz et al. (eds.), *Research and Practice in Chemistry Education*,
https://doi.org/10.1007/978-981-13-6998-8_15

these transferable skills. The tool allows detailed evaluation of the alignment of any assessment item against the claimed set of learning outcomes. We present here two examples in which use of the tool provides evidence for the level of achievement of transferable skills and a third example of use of the tool to inform curriculum design and pedagogy, with the goal of increasing achievement of communication and teamwork bench marks. Implications for practice in assessment design for learning are presented.

Introduction

The importance of undergraduate students acquiring essential transferable skills such as communication, teamwork, problem solving and leadership (sometimes called "soft skills" or "professional skills") has been emphasised in many industry (Confederation of British Industry, 2016; Hager, Holland, & Beckett, 2002), and academic reports (Heckman & Kautz, 2012; Robles, 2012; Sarkar, Overton, Thompson, & Rayner, 2016). Recognising their importance, most institutions in Australia now list a set of graduate attributes (also called graduate outcomes, graduate capabilities and related terms) that a graduating student will have (Barrie, 2004).

Further evidence of the increasing attention being paid to the transferable skills that students should acquire during their degree, is reflected in national and transnational efforts. In Europe, the guidelines for application for the prestigious Eurobachelor label specify a list of generic competences among the required learning outcomes of a degree (European Chemistry Thematic Network, 2016). Similarly, the American Chemical Society's guidelines for evaluation of degree programmes list a set of skills of this type that should be developed during the degree (Committee on Professional Training, 2015). In Australia, the set of Science Threshold Learning Outcomes (TLOs), and the discipline-specific Chemistry Threshold Learning Outcomes (CTLOs) derived from the Learning and Teaching Academic Standards (LTAS) project (Jones, Yates, & Kelder, 2011; Pyke, O'Brien, Yates, & Buntine, 2014) include transferable skills. The CTLOs have been adopted by the Royal Australian Chemical Institute (RACI) for professional accreditation of chemistry degrees in Australia.

Some institutions have implemented specific programmes to target transferable skills, including programmes within science (Peat, Taylor, & Franklin, 2005) and chemistry degrees (Loshbaugh, Laursen, & Thiry, 2011; Towns, 2010). However, the definition and implementation of transferable skills as graduate attributes remain difficult (Green, Hammer, & Star, 2009). Those authors suggest that the attributes should be embedded holistically, through a whole-of-institution approach. An alternative holistic approach focused on the ongoing personal development of the student

I. M. Jamie
Macquarie University, Sydney, NSW, Australia
e-mail: ian.jamie@mq.edu.au

has also emerged (Hill, Walkington, & France, 2016), exemplified by the development of these attributes occurring "through intentional movements [by the student] assisted by institutional arrangements" (Su, 2014). This approach is supported by structured programmes that culminate in well-designed capstone units (Lee & Loton, 2015; Spronken-Smith et al., 2016). As an Australian example, the University of Sydney stated in its Strategic Plan 2016–2020 that a university-wide approach to assessing graduate qualities will be developed (The University of Sydney, 2016). Achievement against these attributes will be reported for every student at the completion of their degrees.

In spite of this recognition of the importance of transferable skills and the efforts listed, it is clear that many students graduate without the ability to apply these skills in the workplace (Confederation of British Industry, 2016; Hager et al., 2002; Sarkar et al., 2016). As Bodner (2016) said in a plenary lecture, "they are called soft skills because they are so hard to develop". Similarly, Knight (2007, p. 1) has referred to these skills as "wicked competences", because "they resist definition, shift shape and are never 'solved'". Thus, further effort is required to develop methods to ensure that students have achieved desired attributes upon graduation.

Assessment has a critical role in the student learning experience (Ramsden, 2003) and can be designed to equip students for life-long learning (Boud & Falchikov, 2006). In addition, it is known that from the students' perspective, assessment defines the course (Gibbs & Simpson, 2004). Thus, it can be argued that if graduate attributes are not assessed, they will not be taken seriously by students (Knight & Page, 2007, p. 11). Following over a decade of work exploring the definition and assessment of graduate attributes (Barrie, Hughes, Crisp, & Bennison, 2012), Barrie and colleagues have argued that embedding these skills within compulsory assessment provides the strongest evidence for their achievement (Hughes & Barrie, 2010). Similarly, because graduate attributes are typically written using ambiguous language, Sadler (2015) argues for the use of assessment to define the expectations of a course. He writes

> …assessment tasks and specifications are material formulations that can be exhibited, argued about and administered. They provide the sharpest and most direct tool available for discussing, clarifying and communicating course intentions for students and academics alike. (p. 5)

Thus, assessment is a starting place to reform the design of teaching, particularly for transferable skills.

In the Australian context, it has been reported that although a majority of academic staff know graduate attributes are important, many do not incorporate these skills into their teaching practice (de la Harpe & David, 2012). While many do provide opportunities for students to gain these skills, they are unlikely to assess them adequately. In a comprehensive report on the assessment of transferable skills, Knight and Page (2007) noted that they can be characterised as "achievements that cannot be neatly pre-specified, take time to develop and resist measurement-based approaches to assessment". Surprisingly, those authors found that many academic staff did not perceive the difficulties in assessing such skills. Improving the ability of academic staff to evaluate the effectiveness of their assessment items could help ameliorate this situation and give them confidence in the design of their assessment items.

The authors constitute the project team for the Office of Learning and Teaching (OLT) funded project "Assessing the Assessments: Evidencing and benchmarking student learning outcomes in Chemistry", OLT 14-3562. This project derives from the development of the Chemistry Threshold Learning Outcomes (CTLOs) (Pyke et al., 2014). Preliminary work on application of the CTLOs to existing programmes of study showed that demonstrating their achievement was not straightforward (Schultz, Mitchell Crow, & O'Brien, 2013). This project seeks to support academic staff by examining the extent to which their assessment measures the intended learning outcomes, and by providing examples of high-quality assessment items. Outcomes from this project align with Sadler's (2015) call for reform in assessment, particularly reform of the design and specification of assessment tasks.

Using intended learning outcomes as the basis to design curriculum, assessment tasks, and learning activities is recognised as optimal practice (Biggs, 1996; Wiggins & McTighe, 2005). Within this approach to instructional design, the first step is to identify the intended learning outcomes, then identify the method of assessment to determine what constitutes acceptable evidence of student achievement of these outcomes. Finally, the learning experiences and instruction are planned to provide students with the opportunity to acquire the knowledge and/or skills that will meet the intended learning outcomes. Application of a verification tool to evaluate the alignment of the assessment items in measuring students' achievement can assist instructors to ensure that those assessment methods are adequate (Fig. 15.1).

An overall goal of the Assessing the Assessments project was to provide academic staff with a tool to diagnose whether a particular assessment item allows students to demonstrate the achievement of a specific learning outcome. The tool can either be applied retrospectively to benchmark assessment, or proactively to inform the planning and design of teaching and assessment. This tool was developed for a specific set of learning outcomes, the CTLOs, but it can be applied to any desired learning outcome or graduate attribute (Schmid et al., 2016). In this chapter, we report the application of the tool to evaluate two specific CTLOs that target transferable

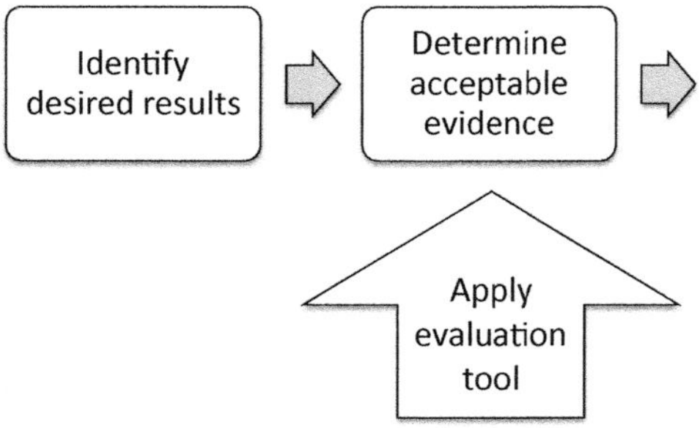

Fig. 15.1 Application of an evaluation tool within the backwards design process

skills: communication (CTLO4.1) and teamwork (CTLO3.5), and illustrate it with three assessment items that have been implemented in upper level chemistry classes.

There are several dimensions in each CTLO expressed as a second tier (Pyke et al., 2014) and we are considering the following specific second tier CTLOs:

3. Investigate and Solve Qualitative and Quantitative Problems in the Chemical Sciences by

3.5: Demonstrating the cooperativity and effectiveness of working in a team environment.

4. Communicate Chemical Knowledge by

4.1: Presenting information, articulating arguments and conclusions, in a variety of modes, to diverse audiences, and for a range of purposes.

These learning outcomes frame the skills that they address specifically in the discipline of chemistry and the chemistry degree. This may lead to differences in outcomes of evaluations of assessment items between applying the CTLOs compared with a set of generic institutional graduate attributes that are not situated within their discipline context.

We have selected CTLOs 3.5 and 4.1 in particular, because teamwork and communication appear ubiquitously in every published list of transferable skills, graduate attributes and generic competences, are cited as highly desired by employers (Sarkar et al., 2016), yet are frequently poorly assessed. Our initial overview mapping of chemistry degrees in Australia in 2012 showed that CTLO4.1 was only addressed in few units (1–3) in a typical chemistry programme (Schultz et al., 2013). That mapping exercise did not evaluate individual assessment items and at that time CTLO3.5 had not been included as a second tier item.

As described in our report of development of the tool, when evaluating submitted assessment items, it was observed that many academic staff assumed that participation in any activities with peers enabled acquisition and assessment of teamwork skills (Schmid et al., 2016). The terms "group work" and "teamwork" are used interchangeably (Johnson & Johnson, 2009; Johnson, Johnson, Ortiz, & Stanne, 1991) so the label alone does not determine whether the item allows students to demonstrate the CTLO. It is therefore important to evaluate assessment of the specific skills gained by students within an activity. Within science faculties in Australia, teamwork skills tend to be poorly taught and rarely assessed and few academic staff are properly trained in this area (Dunne & Rawlins, 2000). Moreover, as described above, both cooperativity within effective teamwork skills and communication skills have been proven difficult to assess via standard assessment methods typically employed in universities (Knight & Page, 2007). Consideration of these two skills together lends the opportunity to design learning activities in which both can be developed and demonstrated. That is, students have the opportunity to work cooperatively in a team for particular tasks, and part of the activity involves some form of communication of their outcomes or findings to others within or beyond the team.

The research questions answered by this study are:

- To what extent can a tool assist academic staff to judge whether their assessment tasks allow students to demonstrate transferable skills, in particular communication and teamwork?

- How can use of a tool assist in the instructional design of assessment methods and aligned learning experiences?

We provide three case studies submitted by members of the project team exemplifying assessment items that allow students to demonstrate achievement of teamwork and communication skills. The overall conduct of the project was approved by the Human Research Ethics Committee of the University of Sydney with approval number 2016/490.

Results and Discussion

The process of designing the tool to evaluate assessment items has been reported (Schmid et al., 2016) and a summary is available at the project website http://chemnet.edu.au/assessment. The outcome of that process is a series of four-square grids for each CTLO plus an assessment rating, shown in Fig. 15.2 and Table 15.1. Each grid in Fig. 15.2 is shaded by the evaluator to reflect the level to which an assessment item confirms a specific learning outcome (either to developing or at graduate level) on the vertical axis, and how much of the stated learning outcome it addresses (partial or whole) on the horizontal axis.

The evaluator then applies an assessment rating according to the extent of the alignment; the levels of attainment are: A (addressed), D (demonstrated) or C (credited, i.e. assessed) as given in Table 15.1. This is a refinement of our previously published version of the tool, based on our experiences evaluating a set of items.

The tool is designed for use with single assessment items (which can then be examined collectively) rather than a suite of assessment items. Nonetheless, evaluation with the tool can be a powerful motivation for change of a set of learning activities and corresponding assessment, as shown through the first example below.

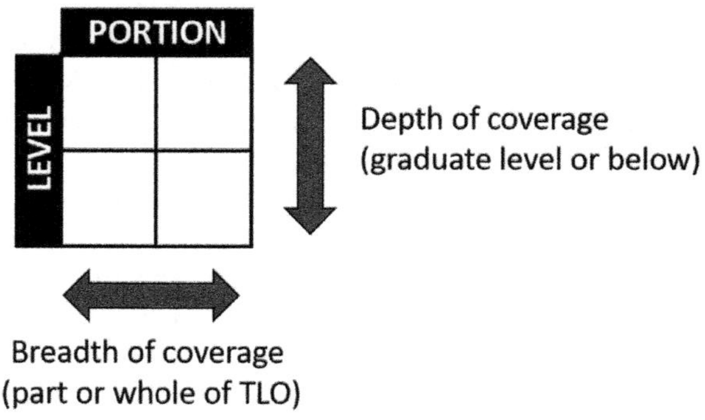

Fig. 15.2 Evaluation tool for an assessment item against a specific learning outcome. Adapted from Schmid et al. (2016)

Table 15.1 Assessment ratings in the tool that describe the level of attainment

Evaluation result		Description
A		The learning outcome is addressed, but students are not required to demonstrate their capability
D		Students are required to demonstrate their capability, but are not credited based on that demonstration
C	(light hatched)	Students are credited based on their demonstrated capability, but a passing grade can be achieved without that credit
	(dark shaded)	Students are credited based on their demonstrated capability, and that credit is a necessary requirement for a passing grade (confirmed)

Application of the Tool

As part of the Assessing the Assessment project, 45 items were submitted by Australian chemistry academic staff for evaluation by the project team. It was found that only eight of the items specifically addressed CTLO3.5 and of these, credit was given for achieving this CTLO in only four items (less than 10% of the submitted assessment items). This indicates a paucity of assessment items providing opportunities for students to develop teamwork skills, and to provide evidence of their growth and achievement. By contrast, 30 of the items addressed and demonstrated CTLO4.1, and 22 of those awarded credit for demonstrating this learning outcome. This shows that communication skills are more frequently addressed; however, there is still a need to ensure that students are given opportunities to communicate in a variety of contexts, including to non-scientific audiences. Figure 15.3 summarises these data.

The full analysis of all 45 submitted tasks is presented in the project website at http://chemnet.edu.au/assessment. Selected exemplars from those tasks are also

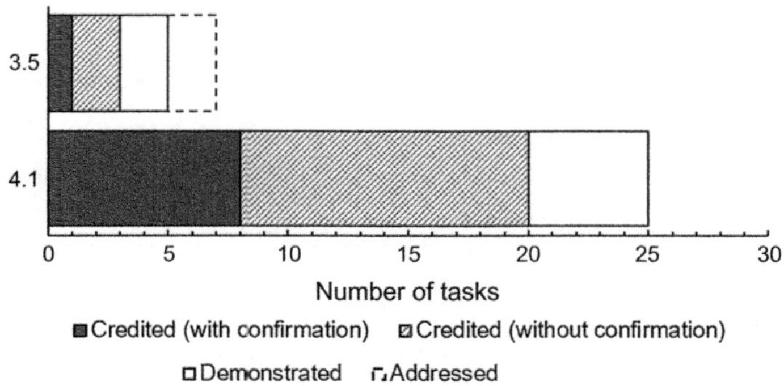

Fig. 15.3 Number of submitted tasks (from a total of 45) that achieved each assessment rating for CTLO3.5 and 4.1

available for adoption, including the three case studies below. All rubrics used in the assessment of the exemplars are available on the website for adoption.

Case Study 1. Team-Based Analytical Laboratory Project

The development of assessment of teamwork and communications skills within a third (final) year unit at an Australian university exemplifies the changed emphasis on these transferable skills in degree programmes over the years. This unit, currently named Chemical Analysis and Inference, constitutes one eighth of an annual full-time load. In this semester-long unit, teams of students plan, execute and report on an investigation of a complex authentic industrial "waste" material (solid or liquid or combination) including background research into the nature of their sample and possible contaminants, and state regulations of disposal options.

Within this laboratory investigation format, the emphasis on the development of effective teamwork skills and communication skills has increased with successive iterations of unit development. Figure 15.4 provides a timeline showing the changes to the structure of the assessment of the laboratory component over the past 20 years. This chart illustrates an increase in the weight and guidance being given to the teamwork and communication aspects of the task. In addition, it shows the change from students working in informal groups in the laboratory prior to 2007 to students carrying out the laboratory investigations as collaborative teamwork. Staff focus on assessment of the success of teamwork developed in parallel with the change in student focus on working with their peers. Such a timeline is not unusual for the development of larger assessment items, where the actual experiments performed do not change but emphases change over time.

Explicit assessment of individual effectiveness within teamwork was introduced from 2007 within the laboratory performance evaluation according to the criteria listed in Table 15.2.

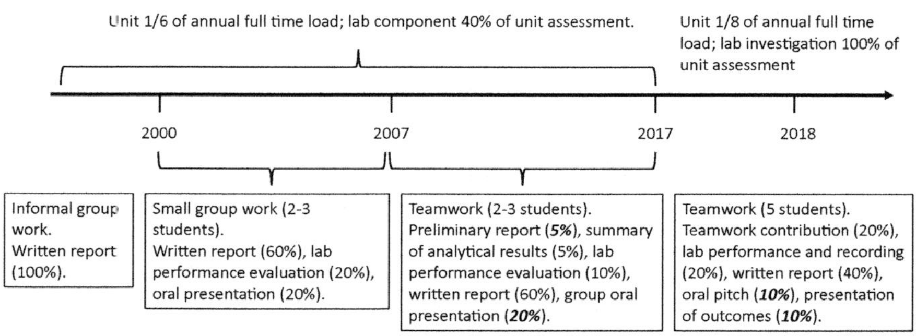

Fig. 15.4 Timeline of assessment of chemical analysis laboratory component showing changes to structure and assessment. Percentages shown in bold italics are team assessment components

Table 15.2 Checklist to scaffold student awareness of the assessment of their teamwork

Lab Performance
Your demonstrators [teaching assistants] will assess your laboratory performance each week. This will involve observation of your activities during the laboratory: • Are you actively involved in experimental work and discussion within your group? • Are you prepared (do you have a clear idea of what you were planning to do? Did you have your risk assessment prepared at the start of the practical time)? Your [laboratory] record book will be checked each week and initialled by a member of staff. The completeness of your record-keeping will also be assessed • Your involvement in the eLearning discussion group for your sample

From 2007 onwards, the students in this unit were carrying out the laboratory investigation as a collaborative team with common goals and shared outcomes (Johnson et al., 1991). The laboratory assessment included both a teamwork component (Table 15.2) and a team communication (presentation) component, both assessed by teaching assistants and academic staff.

The revision of this unit for 2018 arose when final year units changed weight to one eighth of a full-time annual study load. The laboratory project now constitutes the whole of the unit. Although the main components of student activity involve the research and investigation as before, team size and assessment of teamwork and of different forms of communication have been changed. These changes were implemented as a direct result of the concurrent involvement of one academic staff member in developing the tool of the Assessing the Assessment project. This resulted in much increased awareness of the importance of assessing transferable skills.

Team size was increased to five students to provide students with experience of a larger team with a greater diversity of team members, where each finds their own strengths to contribute. Teams now have more time for research, planning and undertaking the chemical analyses. Supporting students in their teams was one of the more important tasks of the teaching team, due to awareness of student issues with teamwork (Wolfe & Powell, 2014). A scaffolded in-class discussion introducing teamwork was held when the teams were formed. This included a brief introduction of Belbin roles type categorisation for team members' self-evaluation (Aritzeta, Swailes, & Senior, 2007; Belbin Associates, 2018; Meslec & Curşeu, 2015; van Dierendonck & Groen, 2011), use of SWOT analysis (strengths, weaknesses, opportunities and threats) along with brief discussion of valuable materials found online (Crebert et al., 2011; OpenLearn, 2018). The teams also discussed and revised the rubric to be used for peer assessment of teamwork contribution, based on rubrics developed within the institution and the literature (Dijkstra, Latijnhouwers, Norbart, & Tio, 2016; Gabelica, Van den Bossche, De Maeyer, Segers, & Gijselaers, 2014; Koh, Hong, & Seah, 2014; Spatar, Penna, Mills, Kutija, & Cooke, 2015). The rubric used is provided in Table 15.3 to indicate the highest and lowest standards for each criteria, because it is straightforward to infer the levels in between these.

Table 15.3 Criteria used in the assessment rubric applied for self- and peer-assessment of teamwork contribution; only the highest and lowest standards are provided for brevity

Category	Excellent 100–85%	Unsatisfactory 50–0%
Intellectual contribution	Contributes scientifically sound ideas and uses wide knowledge of appropriate scientific (chemical) concepts. Initiates possible directions for investigations. Can incorporate and add to others' suggestions. Contributes original and creative ideas for project development	Contributes few scientifically sound ideas. Has scarce valid chemical knowledge to apply. Does not follow up possible directions for investigations. Has little input to project development
Workload	Does a full share of the work; demonstrates initiative. Is aware of others' parts in project, assists others willingly without dominating	Does less work than others. Doesn't get caught up in project after absence; doesn't offer or ask for help
Getting organised	Shows leadership in communication and planning/organisation of team. Completes individual responsibilities and shares with team on time. Attends meetings punctually	Makes little planning contribution. Does not complete some individual responsibilities and contributions are sometime late or compromise quality. Misses some meetings, no reason
Communication among members	Always communicates punctually, openly, clearly and effectively	Rarely communicates punctually, openly, clearly and effectively
Participation in discussions	Can lead discussions, draws out others' contributions. Participates willingly and inspires others. Helps discussion to move forwards	Seems bored with conversations about the project; rarely speaks up and ideas are sometimes off the mark
Supporting a constructive team climate	Treats all team members respectfully, keeps a positive attitude about the team and the project, motivating and encouraging other team members. Values others' contributions. Accepts help. Identifies possible issues and conflicts and helps in team resolution	Does not act to support team morale or progress. Engages in non-committal or negative behaviours. May pursue own tangent and partly disengage
Providing and receiving feedback	Habitually provides helpful, clear, and respectful feedback. Accepts feedback and responds to act on it constructively	Often has little or no feedback to give or feedback is given rudely. Sometimes refuses to listen to feedback or does not act on it

The assessment of teamwork was carried out using four measures as follows:

- students wrote two reflective pieces about their experiences, one in week 3 concerning prior teamwork experiences and one in week 13 (final week of semester) concerning teamwork experiences in this unit;
- staff provided an assessment of student contribution to an online wiki within the learning management system; and
- peer evaluation of individual contribution to teamwork was made using the rubric in Table 15.3. This assessment was carried out at the end of semester following submission of all other assessment items. Because each student is peer assessed by four other team members, averaged peer assessments are less prone to potential bias, which can occur in smaller groups. Peer assessments were anonymous to the students, because they saw only the resultant mark, but not to the staff; and
- self-evaluation of contribution to teamwork was also made using the same rubric at the end of semester. A potential limitation of the use of this assessment protocol for both peer and self-evaluation is that it was carried out at a late time point, and recall of contributions may not have been accurate.

These four measures contributed equally to the overall teamwork assessment, which makes up 20% of the final mark. Student informal feedback on the unit, and especially for the teamwork support and assessment, has been overwhelmingly positive.

Application of the evaluation tool to the different modes of assessment for this laboratory experience over time illustrates how the modifications to assessment protocols result in improved scores on the tool for both CTLO3.5 and 4.1. Figure 15.5 illustrates the outcomes from the tool for these CTLOs for the three earlier assessment modes, and for the full set of CTLOs for the current assessment mode. Note that these outcomes were obtained through evaluations of the assessment item by project team members.

It can be seen that inclusion of assessed oral presentations and collaborative teamwork lead to higher scores using the tool for these CTLOs. In particular, the changes

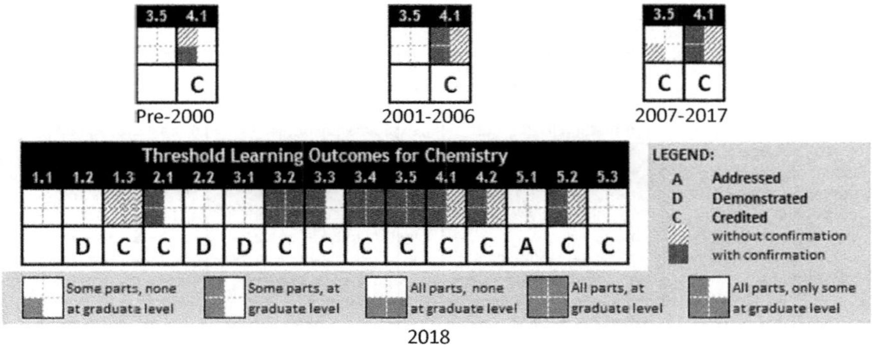

Fig. 15.5 Outcomes from the use of the evaluation tool on the assessments of the analytical laboratory project. Top: outcomes for previous assessment modes. Bottom: outcome for current assessment mode

made for 2018 increased the score for CTLO3.5 from some parts, not at graduate level and without confirmation, to all parts at graduate level with confirmation.

Case Study 2. Supporting Collaborative Group Work in a Nanoscience Course Through a Wiki-Based Laboratory Notebook

The instructional design in this task intentionally aligned assessment of the intended learning outcomes with the learning activities and these were optimised prior to the development of the evaluation tool (Lawrie & Grøndahl, 2015; Lawrie, Grøndahl, Boman, & Andrews, 2016). This case study therefore illustrates an application of the tool to evidence key features of the assessment and how they align with CTLOs 3.5 and 4.1.

The original instructional design aimed to facilitate and evidence the collaborative interactions between individual students using a digital platform (wiki) as they progressed through a multi-week extended experimental investigation. Students worked in groups of three or four and were required to complete all their record-keeping during the stages of the experimental process and communication of their outcomes within their dedicated group wiki. Scaffolding (Fig. 15.6) was introduced to develop students' skills and set expectations for their individual contributions to the group's wiki. The artefacts of this process were formally evaluated along with the effectiveness of different elements of the task in supporting student learning (Lawrie et al., 2016).

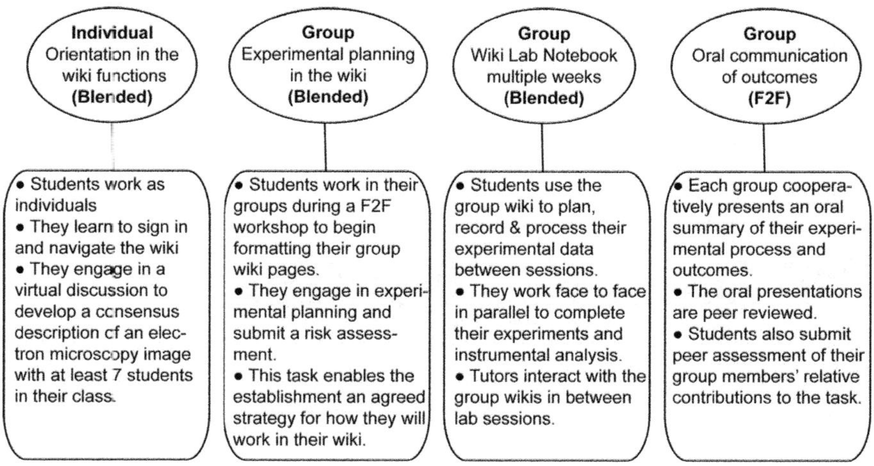

Fig. 15.6 Scaffolding embedded in the task across several weeks to support student engagement and success in the use of a wiki platform for collaboration and communication during the semester. (Blended = combination of online and face-to-face; F2F = face-to-face)

Student activity and achievement in each scaffolded activity was formally assessed. The laboratory component was weighted at 30% of the unit (which is one eighth of an annual full-time load), comprised of 20% for the wiki lab notebook and 10% for an oral presentation. To moderate the wiki lab notebook mark, each individual student's peer assessment of their group members was applied and students completed these anonymously at the end of the laboratory task—their marks were only shared with instructors. The highest and lowest standards applied in the peer assessment criteria provided to students are shown in Table 15.4. There are many similarities to the assessment criteria provided in Table 15.3, illustrating essential aspects of individual contributions to group work that are applicable in many contexts. However, comparison of these two tables also demonstrates the degrees of granularity or elements that can be made explicit in the criteria that evaluate individual student's contributions to collaborative work.

It is important to emphasise that this task represented a blended learning and assessment environment where students collaborated face-to-face during their experimental work in the laboratory sessions as well as asynchronously in the wiki between lab sessions. Many groups also chose to discuss and formulate their wiki entries in the laboratory prior to independent editing. There was strong evidence that this plat-

Table 15.4 Criteria used in the assessment rubric applied for peer assessment to measure individual contribution to group activities; only the highest and lowest standards are provided for brevity

Category	Excellent 100–85%	Unsatisfactory 50–0%
Collaboration	Participated beyond the expectations of the task Showed leadership in setting and meeting goals; maintained group cohesion; and encouraging the best from team members. Showed leadership in the preparation of the e-lab report	Participated minimally Showed little concern for goals Observed but didn't participate in goal setting Completed assigned tasks late or not at all Made no contribution to the e-lab report
Intellectual contribution	Provided an outstanding contribution to the design of experiments, analysis of data and interpretation of results. Demonstrated a deep understanding of the processes	Limited participation in the design of experiments. Did not contribute to analysis of data and interpretation of results. Did not grasp the experimental procedures
Communication	Established and promoted communication networks Encouraged all group members to share their ideas and complete experiments Listened attentively to others and proactively addressed other people's feelings and ideas	Did not share ideas Did not contribute to discussions Did not show consideration for others

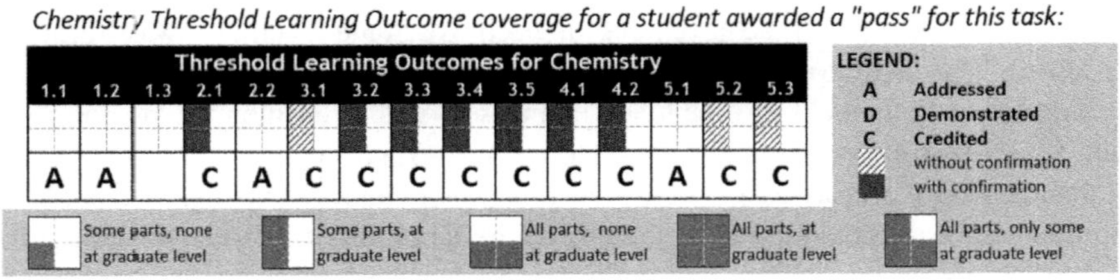

Fig. 15.7 Outcomes of the use of the evaluation tool for the laboratory assessment in nanoscience

form had supported their development of effective cooperative and teamwork skills (Lawrie et al., 2016).

The communication skills assessed in this unit involve the asynchronous discourse in the wiki, the laboratory records, a written discussion of results and an oral presentation. The audience includes their peers, tutors, instructors and instrument scientists. The assessment criteria for the oral presentations include structure, content, design, delivery and professionalism.

This task has been implemented effectively annually since 2010 with minor adjustments based on student feedback. The outcomes of the use of the tool to evaluate this task are shown in Fig. 15.7. Seven CTLOs have been confirmed as being assessed to graduate level (in part) through completion of this task. Hence, the evaluation tool has validated the outcomes of the original instructional design that had targeted learning outcomes described in CTLO3.5 and CTLO4.1, confirming its utility.

Case Study 3. Meta-assessment

The meta-assessment forms 20% of the assessment in a capstone research methods unit, which is core to the chemistry major and comprises one quarter of an annual full-time load. The assessment has been undertaken every year since 2010, with only minor alterations. The three aims to this assessment task are for students to:

- critically evaluate their learning experiences by building a portfolio of assessment tasks from other courses they have studied as part of their undergraduate degree;
- define the outcomes of their assessments within the intended learning outcomes for each course as defined in curricular materials; and,
- align the work and the intended learning outcomes to a CTLO to chart its attainment through the lens of their experiences.

To achieve these goals requires students to meaningfully engage with education research literature on instruction and assessment design, most notably constructive alignment (Biggs, 1999), which outlines a process for co-construction of meaningful assessment (Biggs & Tang, 2007). There are two components within this assessment task:

- an individual reflective journal outlining the student's perspective of intended constructs developed as part of a series of workshops on reflective practice and research on chemistry education;
- a team-based showcase portfolio with embedded artefacts from their prior learning that demonstrate constructive alignment between intended and achieved learning outcomes.

In the individual reflective journals (Boud, 2001), students engage in critical reflection (Mezirow, 1990) and reflexive practice (Bolton, 1999) to draw on their experiences. To assist this process, students appraise theories on teaching and learning of chemistry developed as part of a series of workshops on learning outcomes and constructive alignment; affective domain and sociocultural influences on learning; unique challenges when learning chemistry; and, assessment and evaluation. Students compare methods of research used to evaluate their efficacy in developing conceptual understanding by completing a survey with each workshop that gives them insight into their own perception or skills and develops their understanding of the intended constructs. Students are provided a choice of critical reflection frameworks through supporting materials (Bennett & Evans, 2018).

The showcase portfolio asks students to work in teams of four to explore one of five categories from the CTLO statement, with the outcomes relating to chemistry content knowledge (CTLO2.1) explicitly not allocated. The electronic portfolio is used as a tool for asynchronous team development of the task (Reeves, 2000), which allows for learner engagement in the task (Barrett, 2007). The question posed to students in this portfolio task is: "do you as soon-to-graduate chemists have the evidence to demonstrate your achievement of these outcomes?" The group prepares a showcase of their assessment artefacts, and for each piece of work the group must explain: how the work addresses the intended learning outcome; what their perception of this work was at the time; how their perception has changed now; and, how they will take this achievement on to the next phase of their career (either to further study or into the workforce).

The assessment of this task is undertaken in two parts, worth 5% of the final grade and 15% of the final grade, respectively. First, the entries in the reflective journal are assessed once using a teacher-developed rubric, with criteria including intention, knowledge acquisition, reflection, and impact. Second, the students complete the constructive alignment process as part of the showcase portfolio by developing their own assessment rubric. This rubric is used by students to self-assess their performance, and then used by the teaching staff to do the same. Each self- and teacher-assessment is worth 5% of their grade. The final 5% of the grade comes from anonymous self- and peer-assessment of their teamwork using the Comprehensive Assessment of Team Member Effectiveness rubric (Loughry, Ohland, & Moore, 2007) and system (Ohland et al., 2013).

The primary objective for the assessment of this task was to address two critical, but underrepresented CTLOs in the curriculum. First, to evaluate students' teamwork on co-construction of a task (CTLO3.5) and second, to explicitly evaluate their ability to self-direct their learning and assessment (CTLO5.1). However, the strength of the task is that the evidence collected in the process has the dual purpose of informing curricular developments at the home institution and assuring learning and providing evidence for attainment of every CTLO. Paradoxically, due to the nature of the meta-assessment, the tool does not give any insight in this example because every CTLO is explicitly addressed.

Conclusions and Implications for Practice

The case studies described provide evidence to support the claim that this tool can be applied effectively in the process of instructional design to align assessment of learning outcomes with learning activities. Detailed examination of the evolution of the first example illustrates the typical iterative process of assessment improvement. These exemplars, along with others, are included in the searchable project website (http://chemnet.edu.au/assessment) and we anticipate adoption and adaptation as institutions seek ways to effectively assess transferable skills. In particular, we have shown that teamwork is rarely assessed in chemistry units at Australian institutions and we have provided a mechanism to allow academic staff to improve its coverage within their assessment tasks.

The tool has applicability across disciplines and to any set of intended learning outcomes that address benchmarking by institutions, professional bodies, and other accreditation committees. The tool is available for download from the website.

The experience gained from this project inherently developed expertise in the team and was transformational in itself, resulting in deep reflection on our own assessment tasks.

Acknowledgements The project team acknowledges the collegial engagement in the workshops, activities and outcomes of the Assessing the Assessments project by Australian academics, particularly those who shared their assessment items and course/unit information. We thank Carmel McNaught, who contributed extensively to the project. We acknowledge funding from the Australian Government's Office for Learning and Teaching (OLT ID14-3652).

References

Aritzeta, A., Swailes, S., & Senior, B. (2007). Belbin's team role model: Development, validity and applications for team building. *Journal of Management Studies, 44,* 96–118. https://doi.org/10.1111/j.1467-6486.2007.00666.x.

Barrett, H. C. (2007). Researching electronic portfolios and learner engagement: The REFLECT initiative. *Journal of Adolescent and Adult Literacy, 50,* 436–449. https://doi.org/10.1598/JAAL.50.6.2.

Barrie, S. C. (2004). A research-based approach to generic graduate attributes policy. *Higher Education Research & Development, 23,* 261–275. https://doi.org/10.1080/0729436042000235391.
Barrie, S. C., Hughes, C., Crisp, G., & Bennison, A. (2012). *Assessing and assuring Australian graduate learning outcomes: Principles and practices within and across disciplines.* https://ltr.edu.au/resources/SP10_1879_Barrie_Report%20_2013.pdf. Accessed October 31, 2018.
Belbin Associates. (2018). The nine Belbin team roles. https://www.belbin.com/about/belbin-team-roles/. Accessed October 28, 2018.
Bennett, D., & Evans, C. (2018). Using critical reflection to get the most out of your learning. https://developingemployability.edu.au/tools/learning-experiences-critical-reflection/. Accessed December 2018.
Biggs, J. (1996). Enhancing teaching through constructive alignment. *Higher Education, 32,* 347–364. https://doi.org/10.1007/BF00138871.
Biggs, J. (1999). What the student does: Teaching for enhanced learning. *Higher Education Research & Development, 18,* 57–75. https://doi.org/10.1080/0729436990180105.
Biggs, J., & Tang, C. (2007). *Teaching for quality learning at university: What the student does* (3rd ed.). Maidenhead: SRHE and Open University Press.
Bodner, G. M. (2016). *Introduction to plenary session.* Greeley, CO: Biennial Conference on Chemical Education.
Bolton, G. (1999). Reflections through the looking-glass: The story of a course of writing as a reflexive practitioner. *Teaching in Higher Education, 4,* 193–212. https://doi.org/10.1080/1356251990040204.
Boud, D. (2001). Using journal writing to enhance reflective practice. *New Directions for Adult and Continuing Education, 90,* 9–17. https://doi.org/10.1002/ace.16.
Boud, D., & Falchikov, N. (2006). Aligning assessment with long-term learning. *Assessment & Evaluation in Higher Education, 31,* 399–413. https://doi.org/10.1080/02602930600679050.
Committee on Professional Training. (2015). *Undergraduate professional education in chemistry: ACS guidelines and evaluation procedures for bachelor's degree programs.* American Chemical Society. https://www.acs.org/content/dam/acsorg/about/governance/committees/training/2015-acs-guidelines-for-bachelors-degree-programs.pdf. Accessed October 31, 2018.
Confederation of British Industry. (2016). *The right combination: CBI/Pearson education and skills survey.* http://www.cbi.org.uk/cbi-prod/assets/File/pdf/cbi-education-and-skills-survey2016.pdf. Accessed October 31, 2018.
Crebert, G., Patrick, C.-J., Cragnolini, V., Smith, C., Worsfold, K., & Webb, F. (2011). *Griffith graduate attributes professional skills toolkit.* Brisbane: Griffith University.
de la Harpe, B., & David, C. (2012). Major influences on the teaching and assessment of graduate attributes. *Higher Education Research & Development, 31,* 493–510. https://doi.org/10.1080/07294360.2011.629361.
Dijkstra, J., Latijnhouwers, M., Norbart, A., & Tio, R. A. (2016). Assessing the "I" in group work assessment: State of the art and recommendations for practice. *Medical Teacher*. https://doi.org/10.3109/0142159x.2016.1170796.
Dunne, E., & Rawlins, M. (2000). Bridging the gap between industry and higher education: Training academics to promote student teamwork. *Innovations in Education and Training International, 37,* 361–371. https://doi.org/10.1080/135580000750052973.
European Chemistry Thematic Network. (2016). *Chemistry Eurobachelor label: Guidelines for applications.* http://files.gandi.ws/gandi22757/file/1-eurobachelor_guidelines-1612.pdf. Accessed October 31, 2018.
Gabelica, C., Van den Bossche, P., De Maeyer, S., Segers, M., & Gijselaers, W. (2014). The effect of team feedback and guided reflexivity on team performance change. *Learning and Instruction, 34,* 86–96. https://doi.org/10.1016/j.learninstruc.2014.09.001.
Gibbs, G., & Simpson, C. (2004). Conditions under which assessment supports students' learning. *Learning and Teaching in Higher Education, 2004–05,* 3–31.

Green, W., Hammer, S., & Star, C. (2009). Facing up to the challenge: Why is it so hard to develop graduate attributes? *Higher Education Research & Development, 28,* 17–29. https://doi.org/10.1080/07294360802444339.

Hager, P., Holland, S., & Beckett, D. (2002). *Enhancing the learning and employability of graduates: The role of generic skills*. https://www.bhert.com/publications/position-papers/B-HERTPositionPaper09.pdf. Accessed October 31, 2018.

Heckman, J. J., & Kautz, T. (2012). Hard evidence on soft skills. *Labour Economics, 19,* 451–464. https://doi.org/10.1016/j.labeco.2012.05.014.

Hill, J., Walkington, H., & France, D. (2016). Graduate attributes: Implications for higher education practice and policy. *Journal of Geography in Higher Education, 40,* 155–163. https://doi.org/10.1080/03098265.2016.1154932.

Hughes, C., & Barrie, S. C. (2010). Influences on the assessment of graduate attributes in higher education. *Assessment & Evaluation in Higher Education, 35,* 325–334. https://doi.org/10.1080/02602930903221485.

Johnson, D. W., & Johnson, R. T. (2009). An educational psychology success story: Social interdependence theory and cooperative learning. *Educational Researcher, 38,* 365–379. https://doi.org/10.3102/0013189X09339057.

Johnson, D. W., Johnson, R. T., Ortiz, A. E., & Stanne, M. (1991). The impact of positive goal and resource interdependence on achievement, interaction, and attitudes. *The Journal of General Psychology, 118,* 341–347. https://doi.org/10.1080/00221309.1991.9917795.

Jones, S. M., Yates, B. F., & Kelder, J.-A. (2011). *Science learning and teaching academic standards statement*. https://ltr.edu.au/resources/altc_standards_SCIENCE_240811_v3_0.pdf. Accessed October 31, 2018.

Knight, P. (2007). *Fostering and assessing 'wicked' competences.* http://www.open.ac.uk/opencetl/sites/www.open.ac.uk.opencetl/files/files/ecms/web-content/Knight-(2007)-Fostering-and-assessing-wicked-competences.pdf. Accessed October 31, 2018.

Knight, P., & Page, A. (2007). *The assessment of 'wicked' competences.* http://www.open.ac.uk/opencetl/sites/www.open.ac.uk.opencetl/files/files/ecms/web-content/knight-and-page-(2007)-The-assessment-of-wicked-competences.pdf. Accessed October 31, 2018.

Koh, E., Hong, H., & Seah, J. (2014). *An analytic frame and multi-method approach to measure teamwork competency*. Paper presented at the 14th International Conference on Advanced Learning Technologies.

Lawrie, G. A., & Grøndahl, L. (2015). Wiki technologies and communities: New approaches to assessing individual and collaborative learning in the chemistry laboratory. In J. García-Martínez & E. Serrano-Torresgrosa (Eds.), *Chemistry education: Best practices, opportunities and trends* (pp. 671–691). Weinheim, Germany: Wiley-VCH Verlag GmbH & Co.

Lawrie, G. A., Grøndahl, L., Boman, S., & Andrews, T. (2016). Wiki laboratory notebooks: Supporting student learning in collaborative inquiry-based laboratory experiments. *Journal of Science Education and Technology, 25,* 394–409. https://doi.org/10.1007/s10956-016-9601-0.

Lee, N., & Loton, D. J. (2015). *Capstone curriculum across disciplines: Synthesising theory, practice and policy to provide practical tools for curriculum design*. https://altf.org/wp-content/uploads/2016/08/Lee_N_NSTF_report_2015.pdf. Accessed October 31, 2018.

Loshbaugh, H. G., Laursen, S. L., & Thiry, H. (2011). Reactions to changing times: Trends and tensions in U.S. chemistry graduate education. *Journal of Chemical Education, 88,* 708–715. https://doi.org/10.1021/ed1008574.

Loughry, M. L., Ohland, M. W., & Moore, D. D. (2007). Development of a theory-based assessment of team member effectiveness. *Educational and Psychological Measurement, 67,* 505–524. https://doi.org/10.1177/0013164406292085.

Meslec, N., & Curşeu, P. L. (2015). Are balanced groups better? Belbin roles in collaborative learning groups. *Learning and Individual Differences, 39,* 81–88. https://doi.org/10.1016/j.lindif.2015.03.020.

Mezirow, J. (1990). How critical reflection triggers transformative learning. In J. Mezirow (Ed.), *Fostering critical reflection in adulthood: A guide to transformative and emancipatory learning*. San Francisco, CA: Jossey-Bass.

Ohland, M. W., Loughry, M. L., Woehr, D. J., Bullard, L. G., Felder, R. M., & Finelle, C. J. (2013). The comprehensive assessment of team member effectiveness: Development of a behaviorally anchored rating scale for self- and peer evaluation. *Academy of Management Learning & Education, 11,* 609–630. https://doi.org/10.5465/amle.2010.0177.

OpenLearn (Producer). (2018). Working in groups and teams. https://www.open.edu/openlearn/money-management/management/leadership-and-management/working-groups-and-teams/content-section-0?active-tab=description-tab. Accessed October 26, 2018.

Peat, M., Taylor, C. E., & Franklin, S. (2005). Re-engineering of undergraduate science curricula to emphasise development of lifelong learning skills. *Innovations in Education and Teaching International, 42,* 135–146.

Pyke, S., O'Brien, G., Yates, B., & Buntine, M. (2014). *Chemistry academic standards statement—Revised*. http://chemnet.edu.au/sites/default/files/u39/CHEMISTRY_Academic_Standards_Accreditation_Trial.pdf. Accessed October 31, 2018.

Ramsden, P. (2003). *Learning to teach in higher education* (2nd ed.). London: RoutledgeFalmer.

Reeves, T. C. (2000). Alternative assessment approaches for online learning environments in higher education. *Journal of Educational Computing Research, 23,* 101–111. https://doi.org/10.2190/GYMQ-78FA-WMTX-J06C.

Robles, M. M. (2012). Executive perceptions of the top 10 soft skills needed in today's workplace. *Business Communication Quarterly, 75,* 453–465. https://doi.org/10.1177/1080569912460400.

Sadler, D. R. (2015). Three in-course assessment reforms to improve higher education learning outcomes. *Assessment & Evaluation in Higher Education*. https://doi.org/10.1080/02602938.2015.1064858.

Sarkar, M., Overton, T., Thompson, C., & Rayner, G. (2016). Graduate employability: Views of recent science graduates and employers. *International Journal of Innovation in Science and Mathematics Education, 24,* 31–48.

Schmid, S., Schultz, M., Priest, S. J., O'Brien, G., Pyke, S. M., Bridgeman, A., et al. (2016). Assessing the assessments: Development of a tool to evaluate assessment items in chemistry according to learning outcomes. In M. Schultz, S. Schmid, & T. Holme (Eds.), *Technology and assessment strategies for improving student learning in chemistry* (pp. 225–244). Washington, DC: American Chemical Society.

Schultz, M., Mitchell Crow, J., & O'Brien, G. (2013). Outcomes of the Chemistry Discipline Network mapping exercises: Are the threshold learning outcomes met? *International Journal of Innovation in Science and Mathematics Education, 21,* 81–91.

Spatar, C., Penna, N., Mills, H., Kutija, V., & Cooke, M. (2015). A robust approach for mapping group marks to individual marks using peer assessment. *Assessment & Evaluation in Higher Education, 40,* 371–389. https://doi.org/10.1080/02602938.2014.917270.

Spronken-Smith, R., McLean, A., Smith, N., Bond, C., Jenkins, M., Marshall, S., et al. (2016). A toolkit to implement graduate attributes in geography curricula. *Journal of Geography in Higher Education, 40,* 254–266. https://doi.org/10.1080/03098265.2016.1140129.

Su, Y. (2014). Self-directed, genuine graduate attributes: The person-based approach. *Higher Education Research & Development, 33,* 1208–1220. https://doi.org/10.1080/07294360.2014.911255.

The University of Sydney. (2016). *2016–2020 strategic plan*. Sydney. https://sydney.edu.au/dam/intranet/documents/strategy-and-planning/strategic-plan-2016-20.pdf. Accessed October 29, 2018.

Towns, M. H. (2010). Developing learning objectives and assessment plans at a variety of institutions: Examples and case studies. *Journal of Chemical Education, 87,* 91–96. https://doi.org/10.1021/ed8000039

van Dierendonck, D., & Groen, R. (2011). Belbin revisited: A multitrait–multimethod investigation of a team role instrument. *European Journal of Work and Organizational Psychology, 20,* 345–366. https://doi.org/10.1080/13594321003590580.

Wiggins, G., & McTighe, J. (2005). *Understanding by design* (2nd ed.). Alexandria, VA, USA: Association for Supervision and Curriculum Development.
Wolfe, J., & Powell, E. (2014). *Strategies for dealing with slacker and underperforming teammates in class projects*. Paper presented at the 2014 IEEE International Professional Communication Conference (IPCC), 13–15 October 2014.

Printed by Printforce, the Netherlands